Biomotors

Linear, Rotation, and Revolution Motion Mechanisms

Biomotors
Linear, Rotation, and Revolution Motion Mechanisms

Edited by
Peixuan Guo and Zhengyi Zhao

CRC Press
Taylor & Francis Group
Boca Raton London New York

CRC Press is an imprint of the
Taylor & Francis Group, an **informa** business

CRC Press
Taylor & Francis Group
6000 Broken Sound Parkway NW, Suite 300
Boca Raton, FL 33487-2742

First issued in paperback 2019

ISBN-13: 978-1-4987-0986-6 (hbk)
ISBN-13: 978-0-367-87211-3 (pbk)

Library of Congress Cataloging–in–Publication Data

Names: Guo, Peixuan, author. | Zhao, Zhengyi, author.
Title: Biomotors : linear, rotation, and revolution motion mechanisms /
Peixuan Guo and Zhengyi Zhao.
Description: Boca Raton : Taylor & Francis, 2017. | Includes bibliographical
references and index.
Identifiers: LCCN 2017024716 | ISBN 9781498709866 (hardback : alk. paper)
Subjects: | MESH: Molecular Motor Proteins--chemistry | Molecular Motor
Proteins--physiology | Nanomedicine | Rotation
Classification: LCC QH442 | NLM QU 55 | DDC 572.8--dc23
LC record available at https://lccn.loc.gov/2017024716

Visit the Taylor & Francis Web site at
http://www.taylorandfrancis.com

and the CRC Press Web site at
http://www.crcpress.com

Contents

Concluding remarks and perspectives 93
Competing interests... 95
Glossary... 97
Index 107

Preface

In this book, we focus on the three categories of biomotors and examine their classification, structure, mechanism, and biomedical applications. Biomotors are ingenious nanoscale machines found inside the cells of living things that enable all biological functions; and like all motors, they harness energy to perform a series of repetitive tasks. The way they work is the primary basis of their classification. Revolving motors are present in a wide variety of bacteria, eukaryotic viruses, and dsDNA bacteriophages. Their movement resembles the manner in which the Earth revolves around the Sun. In contrast, rotating motors move in the way that the Earth spins/rotates on its own axis. They include bacterial flagella motors, F_oF_1 ATP synthase (often called F_oF_1 ATPase), and DNA helicases. Linear motors, represented by myosin, kinesin, and dynein, advance in one direction along the cellular track.

Biomotors possess a complex structure with multiple interacting components that are both mobile and stationary, and their framework is highly adapted to their function. Revolving motors, for example, can be distinguished from rotating motors by the diameter/chirality of their central channel. While revolving motors are translocated by two strands of DNA, most rotation motors are translocated by only one strand inside the channel; and as a result, the revolving channel is larger. It is also larger because of the fact that the threads of the channel are not parallel to those of the DNA; whereas they are parallel in rotation motors, allowing for a closer fit. Rotation motors, which bear distinct structural differences among types, function through a nut and bolt mechanism around a central shaft and are generally hexameric. Linear motors have three primary domains, the head, neck, and tail, with the head directing movement, the tail identifying the motor's cargo, and the neck linking the two.

With few exceptions, the main energy source for cellular functions in all biomotors is ATP (adenosine triphosphate), which binds to the substrate and then becomes hydrolyzed leading to movement via conformational changes. Several models exist for dsDNA translocation in revolving biomotors, which vary on the basis of how many parts of the motor are believed to be functionalized with the DNA substrate at any one time.

While the basic structure of rotary motors helicases and F_oF_1 ATPase is similar, their mechanism is not. Helicases unwind two strands of DNA in the same manner that a zipper works; a number of different models exist that explain how the process occurs. F_oF_1 ATPase has two independent motors, F_1 and F_o. F_1 works to rotate the γ subunit with the hydrolysis of one molecule of ATP expended per turn. When F_1 hydrolyzes ATP, F_o pumps a reverse flow of protons through the membrane, transferring from the *a* subunit to the *c*-ring upon one clockwise rotation of the motor. As the name would suggest, linear motors move in a linear fashion, by interacting with a cytoskeletal track. Cycles of ATP binding and hydrolysis promote the forward movement of these motors by causing them to bind to and then release from the protein substrate.

Finally, we examine novel applications of revolving biomotors in the development of highly potent drugs via synthetic versions of viral DNA packaging mechanisms that can directly inject medicines into diseased cells. Research in this area has shown promise in the treatment of conditions that have developed resistance to traditional therapies, such as HIV, hepatitis B, herpes, influenza, and cancer.

Edited by some of the preeminent figures involved in biomotor research, this book is aimed at a variety of readers—from novices who are just learning about biomotors and would like to broaden their knowledge to those who are looking for a concise synopsis. Our approach to these motors is a little less traditional than in the majority of studies: rather than presenting each of them by category (revolving motors and their structure and mechanism followed by that of rotary motors and so on with linear motors), we have chosen to provide an introduction to the common characteristics of the structure and mechanism of typical biomotors from all three categories. A primary objective behind this manner of organizing this book is to examine the similarities and differences among all motors—revolving, rotating, and linear—regardless of type. Our hope is that this book will both serve as a primer on biomotors in general and provide a positive contribution to existing research.

Summary

Biological motors are ubiquitous in living systems. With the recent discovery of widespread revolving biomotors, biological nanomotors are currently classified into three categories: revolving, linear, and rotation motors. Revolving motors are widespread among bacteria, eukaryotic viruses, as well as dsDNA bacteriophages, and possibly including animal and human cells. This book serves as a quick guide for the audience to get a basic understanding of the structure and function of these three types of motors. F_oF_1 ATPase, viral DNA-packaging motors, bacterial chromosome translocases, myosin, kinesin, dynein, flagellar motors, and dsDNA translocases are used as examples to elucidate the relevance of channel size and chirality to the rotating and revolving mechanism; the role of the arginine finger and the asymmetrical hexamer for sequential action on motion direction control; the different stoichiometry of motor structure; the role of ATP in relevance to entropy and conformational alternations for the generation of force; and the advantage of the revolving mechanism to avoid the coiling and tangling of the substrate as compared with the rotation mechanism. Channel chirality is critical in controlling motion direction by one-way traffic: rotation motors use a right-handed channel to drive right-handed dsDNA, whereas revolving motors use a left-handed motor channel to revolve right-handed dsDNA. Rotation motors use a small channel to ensure close contact of the channel wall with the dsDNA bolt; revolving motors have a larger channel to provide room for the bolt to revolve. Binding of ATP to ATPase results in entropy and conformational changes of the ATPase, leading to a high affinity for dsDNA. ATP hydrolysis results in a second entropy and conformational change to lower the DNA affinity, triggering the release of dsDNA for concomitant transfer to the adjacent subunit. The mechanism to control subunit transition and sequential action relies on the "arginine finger" that bridges subunits to sequentially transduce a signal to the adjacent subunit. The appearance of the asymmetrical hexameric structure in many motors is due to the existence of four monomers and one dimer in the hexameric ATPase ring that results from the cross-communication between two adjacent subunits of the ATPase hexamer. Elucidation of motor structure and function and the finding of the asymmetrical hexamer provide clues for why some asymmetrical hexameric ATPase motors have for a long time been reported as a pentameric configuration by cryo-EM since the contact by the arginine finger renders two adjacent ATPase subunits closer than other subunits. Thus, the asymmetrical hexamer would appear as a pentamer by cryo-EM, a technology that acquires the average of many images.

Editors

Peixuan Guo, PhD is currently the Sylvan G. Frank Endowed Chair in Pharmaceutics and Drug Delivery at The Ohio State University College of Pharmacy, and director of the Center for RNA Nanobiotechnology and Nanomedicine at OSU. After serving as a professor and center director at Purdue University for 17 years, he became endowed chair of biomedical engineering at the University of Cincinnati, then an endowed chair of Cancer Nanotechnology and director of the Nanobiotechnology Center at the University of Kentucky before moving to the Ohio State. He is the founding president of the International Society of RNA Nanotechnology and Nanomedicine and founder and director of the Center for RNA Nanobiotechnology and Nanomedicine at OSU.

He earned his PhD in microbiology from the University of Minnesota in 1987 and was an NIH postdoctoral fellow in 1990, assistant professor at Purdue in 1990, tenured in 1993, full professor in 1997, and honored as Purdue Faculty Scholar in 1998. He was also director of the NIH Nanomedicine Development Center from 2006 to 2011. His recognitions include the Pfizer Distinguished Faculty Award in 1995, Lions Club Cancer Research Award in 2006, distinguished alumnus of the University of Minnesota in 2009, and Distinguished Chinese Alumnus of the 100 Years of the University of Minnesota in 2014. He is the editorial board member of five nanotech journals, has reported hundreds of times on television (such as ABC, NBC, and BBC), and has been frequently featured on the websites of the NIH, NSF, MSNBC, and NCI. He is a member of two prominent national nanotech initiatives sponsored by NIST/NIH/NSF and the National Science and Technology Council. He has been a member of the Foreign Examination Panel of the Chinese Academy of Sciences since 2014. He is a cofounder of P&Z Biological Technology, a consultant of Oxford Nanopore Technologies and Nanobio Delivery Pharmaceutical Company.

Peixuan is the world-recognized founder and pioneer of the RNA Nanotechnology field. He constructed the first viral DNA packaging motor *in vitro* (*PNAS*, 1986), discovered the phi29 motor pRNA (*Science*, 1987), assembled infectious dsDNA viruses (*Journal of Virology*, 1995), discovered the pRNA hexamer (*Molecular Cell*, 1998, featured in *Cell*), and pioneered RNA nanotechnology (*Molecular Cell*, 1998; *JNN*, 2003; *Nano Letters*, 2004, 2005; *Nature Nanotechnology* 2010, 2011). His laboratory built a dual imaging system to detect single fluorophores (*EMBO J*, 2007; *RNA*, 2007) and incorporated the phi29 motor channel into a lipid membrane (*Nature Nanotechnology*, 2009) for single-molecule sensing with potential for high throughput dsDNA sequencing. Recently, his laboratory discovered a third class of biomotors that use revolving mechanism without rotation.

Zhengyi Zhao, PhD earned her PhD degree from the University Of Kentucky College of Pharmacy in 2016 under the guidance of Professor Peixuan Guo, one of the world's top scientists in the field of RNA nanotechnology. She earned her bachelor's degree at Shenyang Pharmaceutical University in 2011. She has broad training in molecular biology, nanobiotechnology, biophysics, and pharmaceutics. Her research focuses on the study of the function, mechanism, and application of the bacteriophage phi29 dsDNA packaging nanomotors, and her research work has been reported by many media sources.

Contributors

Venkata Chelikani, PhD, is a microbiologist with extensive experience in biosensors and nanotechnology. He completed his MSc in microbiology from Andhra University, India, in 2005 and lectured undergraduate microbiology courses for 2 years at Aditya College, India. He then went on to complete an MRes in biomedical sciences from the University of Glasgow, United Kingdom. He completed his PhD in biochemistry (developing estrogen-detecting biosensors) from Lincoln University, New Zealand. He worked as a tutor in microbiology both at the University of Canterbury and at Lincoln University for 2 years before carrying out his postdoctoral work on large DNA viruses and their packaging motors at the Indian Institute of Technology (IIT), Mumbai, India. He is currently working as a lab manager in microbiology at Lincoln University, New Zealand. He is interested in working on interdisciplinary projects involving microbiology and nanotechnology.

Ian Grainge, PhD, is currently an Associate Professor at the University of Newcastle, Australia and Australian Research Council Future Fellow. After receiving his BA (1994) and MA (1995) from the University of Cambridge, UK, he earned his PhD at the University of Oxford, UK in 1997. Later, he did postdoctoral research at the University of Texas, Austin; Cancer Research UK, Clare Hall Laboratories; and The University of Oxford.

His research focus is on genomic stability, with an interest in the many mechanisms that organisms employ to keep their genomes intact and pass on that information to subsequent generations. This includes DNA replication, recombination, DNA repair, and chromosome partitioning and segregation. The goal is to understand these processes at the molecular level, and there is a particular fascination with molecular machines, such as helicases and DNA translocases, that underlie many of the vital processes in the cell. Grainge has 20 years of experience in the area of recombination and 15 years of work on molecular motor proteins.

Professor Hiroyuki Noji, PhD, is a Professor in the Department of Applied Chemistry, The University of Tokyo, and is a single-molecule biophysicist. He has been studying the chemomechanical coupling mechanism of F_oF_1 ATP synthase by the use of single-molecule techniques. He is also known as an inventor of the femtoliter chamber array system for single-molecule enzymatic assays that is currently applied in single-molecule digital ELISA. Professor Noji was trained under the supervision of Professor Masasuke Yoshida and earned his PhD from Tokyo Institute of Technology in 1997. After a postdoctoral fellowship in the laboratory of Professor Kazuhiko Kinosita, Jr., he was appointed an Associate Professor at the Institute of Industrial Science, The University of Tokyo in 2001. In 2005, he moved to the Institute of Scientific and Industrial Research, Osaka University as a full professor. Since 2010, he has been a Professor at the Department of Applied Chemistry, The University of Tokyo.

Fengmei Pi, PhD, earned her PhD degree from the University of Kentucky, College of Pharmacy in 2016 under the guidance of Professor Peixuan Guo. She earned her BS and MS from China Pharmaceutical University in 2007. She worked as a research scientist at Deawoong Pharmaceuticals Co. Korea; and senior formulation scientist in China GSK Consumer Healthcare before joining the University of Kentucky in 2012. She has broad training and working experience in pharmaceutical science. Her research focuses on RNA nanotechnology for targeted drug delivery and cancer therapy, and her research work has been reported by media sources.

Darshan Trivedi, PhD, is currently a postdoctoral fellow in the department of Biochemistry at Stanford University. He earned his doctoral degree in physiology from the Pennsylvania State University, College of Medicine in 2014 under the guidance of Dr. Christopher Yengo. He received his BS in biotechnology and MS in biochemistry from India. Darshan has been supported by American Heart Association predoctoral and postdoctoral fellowships during his tenure as a graduate student and now as a postdoctoral fellow. His interests lie at an interface of biochemistry and biophysics of molecular motors. He is specifically interested to understand the molecular mechanism of muscle contraction; its regulation and how small molecules can be targeted to the sarcomere to modulate contractility.

Nancy Wardle, PhD, earned her doctoral degree in French literature from the Ohio State University. She spent many years as a lecturer at Ohio State and other universities. In her contributions to this book, she immensely appreciates the guidance of Dr. Peixuan Guo and his entire research team for their expertise and support.

Christopher M. Yengo, PhD, earned his doctoral degree in molecular physiology and biophysics from The University of Vermont and did his postdoctoral work at The University of Pennsylvania. He was an Assistant Professor in the Department of Biology at The University of North Carolina at Charlotte and is currently a Professor in The Department of Cellular and Molecular Physiology in The College of Medicine at Pennsylvania State University. He is interested in the fundamental mechanism of myosin-based force generation, the function of nonmuscle myosins, regulation of the actin cytoskeleton, and the role of myosin mutations in heart disease and deafness.

Acknowledgments

The editors would like to thank Dr. Maria Spies, University of Iowa, Dan Li, Dr. Kazuhiko Kinosita, Okazaki National Research Institutes; Dr. Toshio Yanagida, Osaka University for critical comments and helpful discussions. The work was supported by NIH grant R01-EB012135, R01-EB019036, TR000875, and U01-CA151648 to PG; NIH grant R01-HL127699 to CMY; NHMRC APP1005697 and ARC Future Fellowship FT120100153 to IG. The content is solely the responsibility of the authors and does not necessarily represent the official views of NIH.

Some of the content of this book, including some text and figures, have previously been published in Guo P, Noji H, Yengo CM, Zhao Z, Grainge I. 2016. Biological nanomotors with a revolution, linear, or rotation motion mechanism. Microbiol Mol Biol Rev 80:161–186. Copyright © 2016, American Society for Microbiology All rights reserved. The editors would like to thank the ASM Press, for kindly allowing the reuse of this material.

chapter one

Classifications and typical examples of biomotors

The nonrotating phenomenon reported in dsDNA packaging motors during recent years (Baumann et al., 2006; Hugel et al., 2007; Chang et al., 2008; Schwartz et al., 2013) stimulated scientists to investigate the motion mechanism of biomotors other than linear or rotation motion types. In 2013, the discovery of the third class of revolving motors in the phi29 dsDNA packaging motor (Schwartz et al., 2013; Zhao et al., 2013) led to the new classifications of biomotors: revolving, linear, and rotation motors (Figure 1.1) (Goldstein and Vale, 1991; Vale, 1993; Hanson et al., 1997; Bukau and Horwich, 1998; Goldman, 1998; Wang et al., 1998; Hirano, 1999; Schwartz et al., 2013; Zhao et al., 2013; Guo, 2014; Guo et al., 2014). Further studies showed that revolving motion was found to be commonly adopted by many dsDNA translocases. In this chapter, the classifications of biomotors and their typical examples are introduced. (Some hexameric dsDNA translocation motors, such as RuvA/RuvB that is involved in DNA homologous recombination, are relatively sophisticated (Rafferty et al., 1996; Han et al., 2006); and some other dsDNA riding motors such as those involved in DNA repair only display two protein subunits. The mechanisms of these motors are more complicated and will not be addressed here.)

1.1 Typical revolving motors

Experimental evidence showing the nonrotation property of these biomotors includes the following: (1) the motor is still active in packaging with the connectors fused to the procapsid protein, making rotation impossible (Baumann et al., 2006), (2) results from single-molecule force spectroscopy combined with polarization spectroscopy studies showed no signals of connector rotation (Hugel et al., 2007), (3) no rotation of the bead clusters, which were tethered to the DNA ends, was observed during active packaging (Chang et al., 2008). DNA was found to twist by as little as 1.5° per base pair translocated (Liu et al., 2014), confirming a nonrotation mechanism since one helical turn of dsDNA is $\sim$10.5 bases and 1.5°/bp $\times$ 10.5 bp/turn = 15.7° is far below 365° per complete helical turn. The way that revolving motors perform work is somewhat similar but distinct to rotating motors in that they have multiple parts, one of which engages in circular motion;

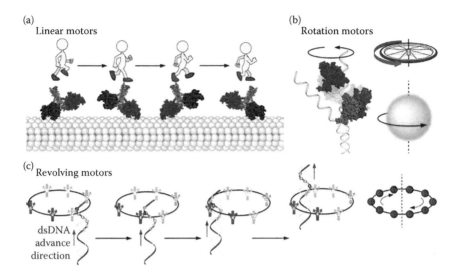

(a) Linear motors

(b) Rotation motors

(c) Revolving motors

dsDNA
advance
direction

Figure 1.1 Illustration of three types of biomotors. (a) Linear motors resembling people walking. (b) Rotation motors like the wheel or Earth rotating on its own axis. (c) Revolving motors like the Earth revolving around the Sun without self-rotation. (Adapted with permission from Guo, P. et al. 2016. *Microbiol Mol Biol Rev.* 80(1), 161–186; and Guo, P. 2014. *Biophys. J.* 106, 1837–1838.)

however, this movement is not akin to how the Earth spins on its own axis, but to how the Earth revolves around the Sun (Figure 1.1) (see animations http://rnanano.osu.edu/movie.html). This characteristic is widespread in a large variety of motors, including dsDNA viruses, dsDNA bacteriophages, and bacteria. Many members of this family are DNA packaging motors or bacterial dsDNA translocases. While there are some structural differences between these biomotors, they share the same basic components, namely a hexameric ATPase ring and a central channel through which DNA passes. The role of this powerful motor in dsDNA viruses is to inject a single piece of DNA into a procapsid protein shell. Bacteriophages are viruses consisting of an infectious tailpiece and a procapsid (head) made of protein. Viruses themselves cannot reproduce on their own; they do so by hijacking host processes to copy their genetic material (DNA or RNA). Because the DNA genome is very long when stretched out, the motor must exert a tremendous amount of force in order to cram the material into the viral capsid; and when released, this pressure forces the DNA into an infected host cell. The injected DNA instructs the bacterium to build proteins, automatically assembling into new capsids. It kills off the bacterium, then attaches itself to other bacteria and repeats the process of infection until every cell has been killed.

Other revolving viral DNA packaging motors include those of T4 phage, said to be twice as powerful as an automobile engine when scaled up, and of herpes virus. The bacterial FtsK (Filamenting temperature-sensitive mutant K) proteins, which comprise a family of motor proteins, also fall into this class of motor. Found in most prokaryotic cells, FtsK is responsible for coordinating the process of copying of the entire complement of genetic material and the late stages of chromosome segregation with cell division.

1.1.1 DNA packaging motor of double-stranded DNA bacteriophages

The revolving mechanism was first proposed in 2013 in the study of the well-researched phi29 dsDNA packaging motor (Schwartz et al., 2013; Zhao et al., 2013). Inspired by the findings in the phi29 biomotor, studies in the other dsDNA motors have been carried out. It was confirmed that the revolving mechanism is a common feature shared by all dsDNA packaging motors, including SPP1, P22, T7, and the HK97 family phage, evidenced by the results from both crystal structural and biochemical studies. During replication, the phi29 bacteriophage translocates its genomic DNA into procapsids (Guo and Lee, 2007; Rao and Feiss, 2008; Serwer, 2010; Zhang et al., 2012), packing DNA against an ever-increasing internal pressure. To overcome this entropically unfavorable process, an energetically favorable motion is required to be coupled to the process of packaging (Guo et al., 1987b; Hwang et al., 1996; Chemla et al., 2005; Sabanayagam et al., 2007). Viral dsDNA packaging motors consist of a dodecameric central channel (Jimenez et al., 1986; Guasch et al., 2002), a hexameric packaging RNA ring (Guo et al., 1987a, 1998; Shu et al., 2007; Zhang et al., 2013), and an ATPase hexameric ring (Figure 1.2). Crystal structure analysis of all the motor channels of SPP1 (Lebedev et al., 2007), T7 (Agirrezabala et al., 2005), HK97 (Juhala et al., 2000), P22 (Olia et al., 2011), and phi29 (Guasch et al., 2002) revealed the existence of a common antichiral arrangement between their channel subunits and the dsDNA helices. It has also been found that in many viruses, dsDNA is naturally spooled inside the viral capsid after packaging, free from rotation tangles (Jiang et al., 2006; Lander et al., 2006; Petrov and Harvey, 2008; Molineux and Panja, 2013). In phi29, a toroid of dsDNA has been shown by Cryo-EM around the portal region (Figure 1.3) (Tang et al., 2008; Sherratt et al., 2010), which is consistent with the revolving mechanism since such toroidal structure images may result from the accumulation of individual revolving DNA processed by Cryo-EM. A compression mechanism found in the T4 DNA packaging motor (Ray et al., 2010; Dixit et al., 2012) also agrees with the revolving mechanism and disagrees with the pentameric gp17 model of T4.

(a)

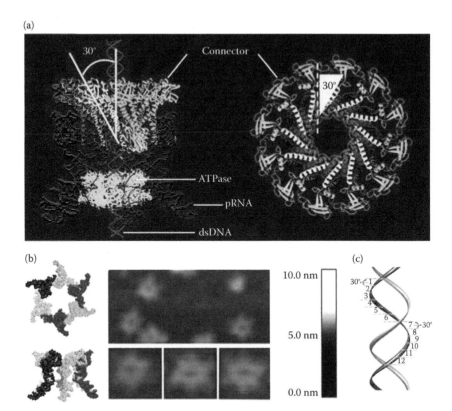

Figure 1.2 Structure of phi29 DNA packaging motor. (a) Side view of phi29 dsDNA packaging motor (left) and top view of phi29 connector (right) based on X-ray crystal structures. (b) Structure of hexameric pRNA in phi29 motor-crystal structure on the left, AFM picture on the right. (c) Representation of DNA contact points whilst revolving inside the connector channel; the contact point with each subsequent connector subunit (see a right hand panel) has a 30° transition step. (Adapted with permission from Schwartz, C. et al., 2013. *Virology* 443, 28–39.)

1.1.2   DNA packaging motor of eukaryotic dsDNA viruses

Adenoviruses (AdV) are a group of well-studied dsDNA viruses that infect eukaryotic cells of vertebrates, including humans. AdV packages its genome with two subunits of the terminal proteins into capsids that contain hexon, penton, and fiber (La et al., 2003). Iva2 and L1 52/55K proteins are AdV packaging proteins. It has been reported that higher-order IVa2-containing complexes formed on adjacent packaging repeats are required for packaging activity (Ostapchuk et al., 2005). Quantitative mass spectrometry, metabolic labeling, and Western blot revealed that there are ∼6–8 IVa2 molecules in each particle (Christensen et al., 2008). The main motor protein Iva2 in AdV is multifunctional. It assists in the assembly

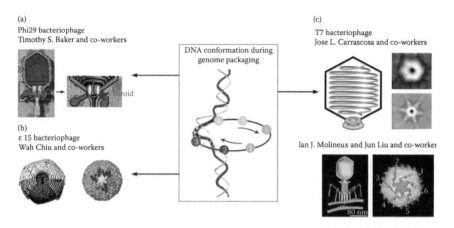

Figure 1.3 Spooling of genome within capsids of various phages. The DNA spooling inside the capsids is shown in (a) phi29 bacteriophage; (b) ε 15 bacteriophage; and (c) T7. The toroid formed at the phi29 portal position as shown in (a) might be a result of accumulation of the images of the revolving motion during packaging. (Adapted with permission from Guo, P. et al., 2014. *Biotechnol. Adv.* 32, 853–872; Jiang, W. et al., 2006. *Nature* 439, 612–616; Cuervo A. et al. 2013. *J Biol Chem.* 288(36), 26290–26299; and Hu, B. et al. 2015. *Proc Natl Acad Sci U S A*. 112(35), E4919–E4928.)

of the capsid and activates late transcription. Comparing the IVa2 protein sequences with ATPase from different species revealed conserved Walker A and B motifs associated with binding and hydrolysis of ATP (Ostapchuk and Hearing, 2008). Similar to the ATPases in bacteriophage packaging motors, the multimeric ATPase IVa2 motor protein complex also works through a sequential action to provide energy for the packaging of DNA into their capsids. IVa2 also interacts with a viral L4-22K protein, which has been shown to be involved in genome encapsidation (Ostapchuk et al., 2006; Ewing et al., 2007; Tyler et al., 2007). IVa2 mutants were defective in DNA packaging and resulted in accumulation of empty capsids similar to the procapsid of dsDNA bacteriophages (Christensen et al., 2008; Ahi et al., 2015).

Herpes simplex viruses (HSV) package their dsDNA genome into a preformed protein shell using terminase (Raoult et al., 2004), which contains a large subunit pUL15 and a small subunit pUL28 (Koslowski et al., 1999). pUL15 cleaves concatemeric viral DNA during packaging initiation and completion cycles and functions as an ATPase, providing energy to the packaging process. X-ray structure of the C-terminal domain of pUL15 showed a homo-trimer structure (Raoult et al., 2004). The structure of the C-terminal domain of pUL15 resembles that of bacteriophage terminases, RNase H, integrases, DNA polymerases, and topoisomerases, with an active site clustered with acidic residues. The DNA-binding surface is surrounded by flexible loops, indicating that they adopt conformational changes upon DNA binding (Selvarajan et al., 2013). These conformational changes are similar to the sequential action of ATPase gp16 observed in phi29 DNA packaging motors, which provides energy to support the one-way traffic of genome into procapsids.

Nucleocytoplasmic large DNA viruses (NCLDV) superfamily include viruses such as Mimivirus, Megavirus, Pandoravirus and Pithovirus (La et al., 2003; Arslan et al., 2011; Philippe et al., 2013) and infect a wide range of eukaryotes (Ghedin and Fraser, 2005; Chelikani et al., 2014b). These viruses are also called giant viruses due to their sheer size (larger than some bacteria). Typical example of these viruses is Mimiviruses which package their 1.2 Mbp dsDNA genome into preformed procapsids through a nonvertex portal (Zauberman et al., 2008) driven by the vaccinia virus A32-type virion packaging ATPase (Monier et al., 2008). It has been shown that the structure and function of their DNA packaging motors are homologous to the FtsK DNA translocase (Iyer et al., 2004; Chelikani et al., 2014b) and use similar revolving mechanism for genome packaging (Iyer et al., 2004; Guo et al., 2016). The genome packaging motors of NCLDVs interact with other genome packaging components such as recombinase and type II topoisomerase similar to prokaryotic FtsK DNA translocase (Iyer et al., 2004; Chelikani et al., 2014a,b). Studies have shown that the FtsK motor operates as a hexamer during genome segregation (Massey et al., 2006) and it is suggested that the hexamer could be a functionally active form of the Mimivirus packaging ATPase (member of NCLDV). The directionality of the FtsK motor movement is provided by the interaction of the γ domain with a short, 8-base-pair DNA sequence known as KOPS (FtsK Orienting Polar Sequence, 5'-GGGNAGGG-3') (Bigot et al., 2005). The γ domain also has a KRKA amino acid loop that is required for the interaction with XerD recombinase (Sivananthan et al., 2009). The Mimivirus packaging ATPase motor also possesses a KRKA motif between residues 227 to 230 toward the C-terminus which could be the potential recombinase interaction site. However, the presence of the KRKA motif is not a conserved feature in NCLDVs. It was also found that potential KOPS-like as well as *dif*-like sequences are present in the Mimivirus genome (Chelikani et al., 2014b). The Mimivirus packaging motor likely gets activated when it encounters a KOPS-like sequence and might recruit topoisomerase II, and this complex is directed to the recombinase already bound at the *dif*-site. The complex so formed resolves the catenated genome and generates an individual unit length of a genome that could still be circular or near circular. Topoisomerase II and recombinase might leave the complex as these proteins could hinder the efficient translocation of the viral genome by packaging ATPase motor.

The capsid assembly and membrane acquisition process leads to the formation of the empty capsid on the periphery of the viral factory. The packaging ATPase still bound to a copy of the resolved genome docks at the membrane–capsid protein interface of the transient opening and the

ATP-dependent translocation of DNA ensues. Finally, after the encapsidation of the whole viral genome, it is proposed that the packaging ATPase leaves the nonvertex packaging site as it has not been identified in viral proteomics analysis (Ronesto et al., 2006). The packaging ATPase can be utilized for another round of packasome complex formation and genome segregation.

Poxviruses, another member of NCLDVs, are large, brick-shaped dsDNA viruses that replicate in the cytoplasm of infected cells. The two DNA strands of the genome are connected at the ends through hairpin termini (Moss, 1985). Poxvirus ATPase is coded by the A32 gene; comparative sequence analysis revealed a highly conserved N-terminal region with five motifs among all poxviruses, including ATPase featuring Walker A and B motifs, A32L-specific motifs III and IV, and a novel motif-V. The secondary structure predictions of N-terminus of A32 ATPase protein are homologous to the FtsK DNA translocase (Yogisharadhya et al., 2012).

1.1.3 DsDNA translocases FtsK/SpoIIIE superfamily

The FtsK family has many varied members, but the majority share the same three components: a N-terminal domain with four transmembrane segments that anchor the motor to the cytoplasmic membrane, a C-terminal domain consisting of about 512 amino acids, and a highly variable linker that connects the N- and C-domains and is rich in amino acids glutamine and proline. While the N-terminal domain plays a major role in cell division, the C-terminal domain functions in chromosome segregation and DNA translocation. The latter is divided into three segments: α, β, and γ. The first two parts are the motor portion of FtsK, and they assemble into a hexameric ring with a central channel through which dsDNA passes. The γ subdomain provides for directionality of the DNA. It has been found that the C-terminal domain of FtsK uses the γ subdomain to bind to DNA at sites of 8 base pair sequences called KOPS, which direct the rapid translocation of DNA. As with many other biomotors, it is the hydrolysis of ATP that powers this process, allowing FtsK to translocate up to 17,000 bp/sec (Lee et al., 2012). FtsK is often grouped in the same superfamily as SpoIIIE, and members of this large family are involved in DNA conjugation, segregation, and translocation as well as protein transport. The FtsK/SpoIIIE family of proteins belongs to the additional strand conserved E (ASCE) superfamily, which are hexameric dsDNA translocases found in many bacterial species. Within the ASCE class, the FtsK/HerA clade is present throughout bacteria and archaea. The large FtsK-HerA family (Iyer et al., 2004) also contains the motor proteins of various conjugative plasmids and transposons such as the single-strand translocase protein TrwB (Gomis-Ruth and Coll, 2001; Gomis-Ruth et al., 2001).

1.2   *Typical rotary motors*

Rotation biomotors share a similar characteristic of movement: as the name would suggest, they work via rotation, in the same way that the Earth spins on its own axis. The category of rotation motors comprises a large number of classes, including bacterial flagella motors, helicases, and the F_oF_1-ATP synthase family of proteins (or F_oF_1-ATPase for short) (Junge et al., 1997; DeRosier, 1998; Enemark and Joshua-Tor, 2006). Bacterial flagella motors lie at the base of flagellar filaments and allow a bacterium to "swim." F_oF_1-ATPase plays an indispensable role by converting proton gradients into ATP, which powers nearly all cellular processes. Helicases separate the two strands of the double-stranded DNA helix into single strands, making it possible for them to be replicated, in addition to many other functions.

1.2.1 F_oF_1 *complex*

F_oF_1 ATP synthase, as an ATP generator, is located in the inner mitochondria membranes, the chloroplasts thylakoid membranes, and the bacteria plasma membranes (Yoshida et al., 2003; Okuno et al., 2011). Its two components, rotary motors F_o and F_1, are connected by a common rotor axle and peripheral stalk (Figure 1.4). This typical rotary molecular energy converter interconverts proton motive force (pmf) and energy of ATP through the rotation of its subunit complex (1–3) (Hiroyuki and Masasuke, 2001).

1.2.2   *DNA helicase*

DNA helicases are a primary class of enzymes in genome maintenance. They play a vital role in cell biology by breaking the hydrogen bonds between the double helix base pairing of double-stranded DNA, allowing the individual strands to be copied, repaired, and recombined. When they are copied, new strands of DNA can be synthesized by DNA polymerase. Helicases are so ubiquitous that mutations or dysfunctions can cause a number of seemingly unrelated diseases and conditions, including cancer, immunodeficiency, and premature aging. The energy source for these biomotors is derived from the hydrolysis of NTP, typically ATP. Upon binding with DNA, the motor's structure is reinforced and NTPase activity is stimulated. Most helicases are hexameric, though some can be dimeric or monomeric. Much like the F_1 motor of F_oF_1-ATPase, the hexamer forms a ring shape with a central channel. What moves through this channel is a single strand of dsDNA, where the DNA strand serves as the stator and the helicase the rotor of the typical rotating motor. The bond between the strand and the channel is very close, and the helicase easily slides over the strand because both are right-handed and their threads

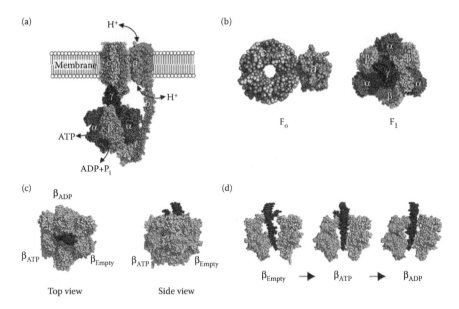

Figure 1.4 Structure of $F_o F_1$ ATP synthase and $\alpha_3 \beta_3 \gamma$ subcomplex of F_1. (a) Reconstituted structure of $F_o F_1$ ATP synthase from crystal structures of isolated subunit or subcomplexes; $\alpha_3 \beta_3 \gamma \epsilon$ subcomplex (PDB code; 3OAA), δ (PDB code; 1ABV), *b* dimer (PDB code; 1B9U, 2KHK, 1L2P), *c*-ring (PDB code; 3UD0), and putative structure of the *a* subunit (PDB code; 1C17). Greenish parts represent the stator complex including the peripheral stalk (δ-b_2 subcomplex) that holds the $\alpha_3 \beta_3$ stator ring of F_1 and ab_2 stator of F_o. Brownish parts represent the rotor complex ($\gamma \epsilon$-*c*-ring subcomplex). (b) F_o and F_1 isolated from Figure 1.1a, both viewed from the top. (c) Crystal structure of F_1 from bovine mitochondria (PDB code; 1BMF). (d) Side views of the conformational states of 3β subunits. β_{Empty} has an open conformation in which the α-helical C-terminal domain rotates upward, opening the cleft of the nucleotide-binding pocket. Both β_{ATP} and β_{ADP} have a closed conformation entrapping the nucleotide within the closed pocket. All α subunits represent the open conformation. α, β, and γ subunits are shown in yellow, green, and red spheres, respectively. (Adapted with permission from Guo, P. et al. 2016. *Microbiol Mol Biol Rev.* 80(1), 161–186.)

parallel, in the same way that a screw passes through a bolt when it is turned.

Nucleic acid helicases can be classified through numerous ways. On the basis of amino acid sequence comparisons, helicases are divided into three large superfamilies (F1, F2, and F3) and two smaller families (F4 and F5). DNA helicases and RNA helicases can both be found in each of these helicase superfamilies except for F6 (Jankowsky and Fairman-Williams, 2010). F5 contains a bacterial Rho factor involved in transcription termination regulation, and F6 contains a structure that functions similarly to the

Rho factor but is different in structure (Jankowsky, 2010). The cores of F1 and F2 enzymes are structurally homologous and contain similar sets of seven helicase signature motifs.

1.2.3 Bacterial flagella

Flagella, with a still helical-shaped structure, are the major force generators for most motile bacteria. Each flagella filament has an individual rotation motor at its base (DeRosier, 1998; Berry, 2001), which can rotate clockwise to generate forward movement, or counterclockwise for tumbling or direction alternation. The flagellar motor is powered by pmf with ATP as an energy source. A number of models have been proposed to explain the mechanism of flagellar motor function (Khan and Berg, 1983; Berg and Turner, 1993; Berry, 1993; Elston and Oster, 1997; Lauger, 1988; Berry and Berg, 1999; Thomas et al., 1999; Walz and Caplan, 2000; Atsumi, 2001), for example, the "electrostatic proton turbine" model (Lauger, 1988; Berry, 1993; Elston and Oster, 1997) and the "turnstile" model (Khan and Berg, 1983; Berg and Turner, 1993). Since many excellent updated reviews are available, readers are encouraged to refer to Guo et al. (2014, 2013), Stevenson et al. (2015), Dutcher (2014), Fisch and Dupuis-Williams (2011), Van et al. (2011); Inaba (2011), and Minamino et al. (2008).

1.3 Typical linear motors

Linear motors were the first described motor proteins. Unlike the rotation mechanism of the previous class of motors, they advance unidirectionally along a cytoskeletal track. This category includes three main types of motors—myosin, kinesin, and dynein—each of which has a large number of subtypes. Like the majority of other motors, they depend on ATP to produce mechanical work. They are responsible for most of the movement in eukaryotic cells. Myosin, for example, plays a primary role in muscle contraction as well as a variety of other intracellular functions. Kinesin is involved in several activities including meiosis, mitosis, and intracellular transport of cellular cargo.

References

Agirrezabala, X., Martin-Benito, J., Valle, M., Gonzalez, J. M., Valencia, A., Valpuesta, J. M., Carrascosa, J. L. 2005. Structure of the connector of bacteriophage T7 at 8A resolution: Structural homologies of a basic component of a DNA translocating machinery. *J. Mol. Biol.* 347, 895–902.

Ahi, Y. S. et al. 2015. Adenoviral L4 33K forms ring-like oligomers and stimulates ATPase activity of IVa2: Implications in viral genome packaging. *Front Microbiol.* 6, 318.

Arslan, D., Legendre, M., Seltzer, V., Abergel, C., Claverie, J. M. 2011. Distant Mimivirus relative with a larger genome highlights the fundamental features of Megaviridae. *Proc. Natl. Acad. Sci. USA* 108, 17486–17491.

Atsumi, T. 2001. An ultrasonic motor model for bacterial flagellar motors. *J. Theor. Biol.* 213, 31–51.

Baumann, R. G., Mullaney, J., Black, L. W. 2006. Portal fusion protein constraints on function in DNA packaging of bacteriophage T4. *Mol. Microbiol.* 61, 16–32.

Berg, H., Turner, L. 1993. Torque generated by the flagellar motor of *Escherichia coli*. *Biophys. J.* 65, 2201–2219.

Berry, R. 1993. Torque and switching in the bacterial flagellar motor. An electrostatic model. *Biophys. J.* 64, 961–973.

Berry, R., Berg, H. 1999. Torque generated by the flagellar motor of *Escherichia coli* while driven backward. *Biophys. J.* 76, 580–587.

Berry, R. M. 2001. Bacterial flagella: Flagellar motor. In: *Encyclopedia of Life Sciences.* John Wiley & Sons, Ltd., Chichester, UK.

Bigot, S., Saleh, O. A., Lesterlin, C., Pages, C., El Karoui, M., Dennis, C., Grigoriev, M., Allemand, J. F., Barre, F. X., Cornet, F. 2005. KOPS: DNA motifs that control E. coli chromosome segregation by orienting the FtsK translocase. *The EMBO Journal* 24(21), 3770–3780.

Bukau, B., Horwich, A. L. 1998. The Hsp70 and Hsp60 chaperone machines. *Cell* 92, 351–366.

Chang, C., Zhang, H., Shu, D., Guo, P., Savran, C. 2008. Bright-field analysis of phi29 DNA packaging motor using a magnetomechanical system. *Appl. Phys. Lett.* 93, 153902–153903.

Chelikani, V., Ranjan, T., Kondabagil, K. 2014a. Revisiting the genome packaging in viruses with lessons from the "Giants". *Virology* 466–467, 15–26.

Chelikani, V., Ranjan, T., Zade, A., Shukla, A., Kondabagil, K. 2014b. Genome segregation and packaging machinery in *Acanthamoeba polyphaga* mimivirus is reminiscent of bacterial apparatus. *J. Virol.* 88, 6069–6075.

Chemla, Y. R., Aathavan, K., Michaelis, J., Grimes, S., Jardine, P. J., Anderson, D. L., Bustamante, C. 2005. Mechanism of force generation of a viral DNA packaging motor. *Cell* 122, 683–692.

Christensen, J. B., Byrd, S. A., Walker, A. K., Strahler, J. R., Andrews, P. C., Imperiale, M. J. 2008. Presence of the adenovirus IVa2 protein at a single vertex of the mature virion. *J. Virol.* 82, 9086–9093.

Cuervo A. et al. 2013. Structural characterization of the bacteriophage T7 tail machinery. *J Biol Chem.* 288(36), 26290–26299.

Deng, F., Wang, R., Fang, M., Jiang, Y., Xu, X., Wang, H., Chen, X. et al. 2007. Proteomics analysis of *Helicoverpa armigera* single nucleocapsid nucleopolyhedrovirus identified two new occlusion-derived virus-associated proteins, HA44 and HA100. *J. Virol.* 81, 9377–9385.

DeRosier, D. J. 1998. The turn of the screw: The bacterial flagellar motor. *Cell* 93, 17–20.

Dixit, A. B., Ray, K., Black, L. W. 2012. Compression of the DNA substrate by a viral packaging motor is supported by removal of intercalating dye during translocation. *Proc. Natl. Acad. Sci. USA* 109, 20419–20424.

Dutcher, S. K. 2014. The awesome power of dikaryons for studying flagella and basal bodies in *Chlamydomonas reinhardtii*. *Cytoskeleton (Hoboken.)* 71, 79–94.

Elston, T., Oster, G. 1997. Protein turbines. I: The bacterial flagellar motor. *Biophys. J.* 73, 703–721.

Enemark, E. J., Joshua-Tor, L. 2006. Mechanism of DNA translocation in a replicative hexameric helicase. *Nature* 442, 270–275.

Ewing, S. G., Byrd, S. A., Christensen, J. B., Tyler, R. E., Imperiale, M. J. 2007. Ternary complex formation on the adenovirus packaging sequence by the IVa2 and L4 22-kilodalton proteins. *J. Virol.* 81, 12450–12457.

Fisch, C., Dupuis-Williams, P. 2011. [The rebirth of the ultrastructure of cilia and flagella]. *Biol. Aujourdhui.* 205, 245–267.

Ghedin, E., Fraser, C. M. 2005. A virus with big ambitions. *Trends Microbiol.* 13, 56–57.

Goldman, Y. E. 1998. Wag the tail: Structural dynamics of actomyosin. *Cell* 93, 1–4.

Goldstein, L., Vale, R. 1991. Motor proteins. A brave new world for dynein. *Nature* 352, 569–570.

Gomis-Ruth, F. X., Coll, M. 2001. Structure of TrwB, a gatekeeper in bacterial conjugation. *Int. J. Biochem. Cell Biol.* 33, 839–843.

Gomis-Ruth, F. X., Moncalian, G., Perez-Luque, R., Gonzalez, A., Cabezon, E., de la, C. F., Coll, M. 2001. The bacterial conjugation protein TrwB resembles ring helicases and F1-ATPase. *Nature* 409, 637–641.

Guasch, A., Pous, J., Ibarra, B., Gomis-Ruth, F. X., Valpuesta, J. M., Sousa, N., Carrascosa, J. L., Coll, M. 2002. Detailed architecture of a DNA translocating machine: The high-resolution structure of the bacteriophage phi29 connector particle. *J. Mol. Biol.* 315, 663–676.

Guo, P. 2014. Biophysical studies reveal new evidence for one-way revolution mechanism of bacteriophage phi29 DNA packaging motor. *Biophys. J.* 106, 1837–1838.

Guo, P., Erickson, S., Anderson, D. 1987a. A small viral RNA is required for *in vitro* packaging of bacteriophage phi29 DNA. *Science* 236, 690–694.

Guo P., Noji, H., Yengo, C. M., Zhao, Z., Grainge, I. 2016. Biological nanomotors with a revolution, linear, or rotation motion mechanism. *Microbiol Mol Biol Rev.* 80(1), 161–186.

Guo, P., Peterson, C., Anderson, D. 1987b. Prohead and DNA-gp3-dependent ATPase activity of the DNA packaging protein gp16 of bacteriophage phi29. *J. Mol. Biol.* 197, 229–236.

Guo, P., Schwartz, C., Haak, J., Zhao, Z. 2013. Discovery of a new motion mechanism of biomotors similar to the earth revolving around the sun without rotation. *Virology* 446, 133–143.

Guo, P., Zhang, C., Chen, C., Trottier, M., Garver, K. 1998. Inter-RNA interaction of phage phi29 pRNA to form a hexameric complex for viral DNA transportation. *Mol. Cell* 2, 149–155.

Guo, P., Zhao, Z., Haak, J., Wang, S., Wu, D., Meng, B., Weitao, T. 2014. Common mechanisms of DNA translocation motors in bacteria and viruses using one-way revolution mechanism without rotation. *Biotechnol. Adv.* 32, 853–872.

Guo, P. X., Lee, T. J. 2007. Viral nanomotors for packaging of dsDNA and dsRNA. *Mol. Microbiol.* 64, 886–903.

Han, Y. W., Tani, T., Hayashi, M., Hishida, T., Iwasaki, H., Shinagawa, H., Harada, Y. 2006. Direct observation of DNA rotation during branch migration of Holliday junction DNA by *Escherichia coli* RuvA-RuvB protein complex. *Proc. Natl. Acad. Sci. USA* 103, 11544–11548.

Hanson, P. I., Roth, R., Morisaki, H., Jahn, R., Heuser, J. E. 1997. Structure and conformational changes in NSF and its membrane receptor complexes visualized by quick-freeze/deep-etch electron microscopy. *Cell* 90, 523–535.

Hirano, T. 1999. SMC-mediated chromosome mechanics: A conserved scheme from bacteria to vertebrates? *Genes Dev.* 13, 11–19.

Hiroyuki, N., Masasuke, Y. 2001. The rotary machine in the cell, ATP synthase. *J. Biol. Chem.* 276(3), 1665–1668.

Hoffmann, C. J., Thio, C. L. 2007. Clinical implications of HIV and hepatitis B co-infection in Asia and Africa. *Lancet Infect. Dis.* 7, 402–409.

Hu, B. et al. 2015. Structural remodeling of bacteriophage T4 and host membranes during infection initiation. *Proc Natl Acad Sci U S A.* 112(35), E4919–E4928.

Huang, N., Wu, W., Yang, K., Passareli, A. L., Rhormann, G. F., Clem, R. J. 2011. Baculovirus infection induces a DNA damage response that is required for efficient viral replication. *J. Virol.* 85, 12547–12556.

Hugel, T., Michaelis, J., Hetherington, C. L., Jardine, P. J., Grimes, S., Walter, J. M., Faik, W., Anderson, D. L., Bustamante, C. 2007. Experimental test of connector rotation during DNA packaging into bacteriophage phi29 capsids. *PLoS Biol.* 5, 558–567.

Hwang, Y., Catalano, C. E., Feiss, M. 1996. Kinetic and mutational dissection of the two ATPase activities of terminase, the DNA packaging enzyme of bacteriophage lambda. *Biochemistry* 35, 2796–2803.

Inaba, K. 2011. Sperm flagella: Comparative and phylogenetic perspectives of protein components. *Mol. Hum. Reprod.* 17, 524–538.

Iyer, L. M., Makarova, K. S., Koonin, E. V., Aravind, L. 2004. Comparative genomics of the FtsK-HerA superfamily of pumping ATPases: Implications for the origins of chromosome segregation, cell division and viral capsid packaging. *Nucleic Acids Res.* 32, 5260–5279.

Jankowsky, E. 2010. *RNA Helicases.* Royal Society of Chemistry.

Jankowsky, E., Fairman-Williams, M. E. 2010. An introduction to RNA helicases: Superfamilies, families, and major themes. In: E. Jankowsky (Ed.), *RNA Helicases.* Royal Society of Chemistry, UK, pp. 1–31.

Jiang, W., Chang, J., Jakana, J., Weigele, P., King, J., Chiu, W. 2006. Structure of epsilon15 bacteriophage reveals genome organization and DNA packaging/injection apparatus. *Nature* 439, 612–616.

Jimenez, J., Santisteban, A., Carazo, J. M., Carrascosa, J. L. 1986. Computer graphic display method for visualizing three-dimensional biological structures. *Science* 232, 1113–1115.

Juhala, R. J., Ford, M. E., Duda, R. L., Youlton, A., Hatfull, G. F., Hendrix, R. W. 2000. Genomic sequences of bacteriophages HK97 and HK022: Pervasive genetic mosaicism in the lambdoid bacteriophages. *J. Mol. Biol.* 299, 27–51.

Junge, W., Lill, H., Engelbrecht, S. 1997. ATP synthase: An electrochemical transducer with rotatory mechanics. *Trends Biochem. Sci.* 22, 420–423.

Khan, S., Berg, H. 1983. Isotope and thermal effects in chemiosmotic coupling to the flagellar motor of Streptococcus. *Cell* 32, 913–919.

Koslowski, K. M., Shaver, P. R., Casey, J. T., Wilson, T., Yamanaka, G., Sheaffer, A. K., Tenney, D. J., Pederson, N. E. 1999. Physical and functional interactions between the herpes simplex virus UL15 and UL28 DNA cleavage and packaging proteins. *J. Virol.* 73, 1704–1707.

La, S. B., Audic, S., Robert, C., Jungang, L., de, L. X., Drancourt, M., Birtles, R., Claverie, J. M., Raoult, D. 2003. A giant virus in amoebae. *Science* 299, 2033.

Lander, G. C., Tang, L., Casjens, S. R., Gilcrease, E. B., Prevelige, P., Poliakov, A., Potter, C. S., Carragher, B., Johnson, J. E. 2006. The structure of an infectious P22 virion shows the signal for headful DNA packaging. *Science* 312, 1791–1795.

Lauger, P. 1988. Torque and rotation rate of the bacterial flagellar motor. *Biophys. J.* 53, 53–65.

Lebedev, A. A., Krause, M. H., Isidro, A. L., Vagin, A. A., Orlova, E. V., Turner, J., Dodson, E. J., Tavares, P., Antson, A. A. 2007. Structural framework for DNA translocation via the viral portal protein. *EMBO J.* 26, 1984–1994.

Lee, J. Y. et al. 2012. Single-molecule imaging of DNA curtains reveals mechanisms of KOPS sequence targeting by the DNA translocase FtsK. *Proc Natl Acad Sci U S A.* 109, 6531–6536.

Liu, S., Chistol, G., Hetherington, C. L., Tafoya, S., Aathavan, K., Schnitzbauer, J., Grimes, S., Jardine, P. J., Bustamante, C. 2014. A viral packaging motor varies its DNA rotation and step size to preserve subunit coordination as the capsid fills. *Cell* 157, 702–713.

Massey, T. H., Mercogliano, C. P., Yates, J., Sherratt, D. J., and Löwe, J. 2006. Double-stranded DNA translocation: Structure and mechanism of hexameric FtsK. *Molecular cell* 23(4): 457–469.

Minamino, T., Imada, K., Namba, K. 2008. Molecular motors of the bacterial flagella. *Curr. Opin. Struct. Biol.* 18, 693–701.

Molineux, I. J., Panja, D. 2013. Popping the cork: Mechanisms of phage genome ejection. *Nat. Rev. Microbiol.* 11, 194–204.

Monier, A., Larsen, J. B., Sandaa, R. A., Bratbak, G., Claverie, J. M., Ogata, H. 2008. Marine mimivirus relatives are probably large algal viruses. *Virol. J.* 5, 12.

Moss, B. 1985. Replication of poxviruses. In: B. N. Fields et al. (Eds.), *Virology.* Raven Press, New York, pp. 685–703.

Okuno, D., Iino, R., Noji, H. 2011. Rotation and structure of F_oF_1-ATP synthase. *J. Biochem.* 149, 655–664.

Olia, A. S., Prevelige, P. E., Johnson, J. E., Cingolani, G. 2011. Three-dimensional structure of a viral genome-delivery portal vertex. *Nat. Struct. Mol. Biol.* 18, 597–603.

Ostapchuk, P., Anderson, M. E., Chandrasekhar, S., Hearing, P. 2006. The L4 22-kilodalton protein plays a role in packaging of the adenovirus genome. *J. Virol.* 80, 6973–6981.

Ostapchuk, P., Hearing, P. 2008. Adenovirus IVa2 protein binds ATP. *J. Virol.* 82, 10290–10294.

Ostapchuk, P., Yang, J., Auffarth, E., Hearing, P. 2005. Functional interaction of the adenovirus IVa2 protein with adenovirus type 5 packaging sequences. *J. Virol.* 79, 2831–2838.

Petrov, A. S., Harvey, S. C. 2008. Packaging double-helical DNA into viral capsids: Structures, forces, and energetics. *Biophys. J.* 95, 497–502.

Philippe, N., Legendre, M., Doutre, G., Coute, Y., Poirot, O., Lescot, M., Arslan, D. et al. 2013. Pandoraviruses: Amoeba viruses with genomes up to 2.5 Mb reaching that of parasitic eukaryotes. *Science* 341, 281–286.

Rafferty, J. B., Sedelnikova, S. E., Hargreaves, D., Artymiuk, P. J., Baker, P. J., Sharples, G. J., Mahdi, A. A., Lloyd, R. G., Rice, D. W. 1996. Crystal structure of DNA recombination protein RuvA and a model for its binding to the Holliday junction. *Science* 274, 415–421.

Rao, V. B., Feiss, M. 2008. The bacteriophage DNA packaging motor. *Annu. Rev. Genet.* 42, 647–681.

Raoult, D., Audic, S., Robert, C., Abergel, C., Renesto, P., Ogata, H., La, S. B., Suzan, M., Claverie, J. M. 2004. The 1.2-megabase genome sequence of Mimivirus. *Science* 306, 1344–1350.

Raoult, D., La, S. B., Birtles, R. 2007. The discovery and characterization of Mimivirus, the largest known virus and putative pneumonia agent. *Clin. Infect. Dis.* 45, 95–102.

Ray, K., Sabanayagam, C. R., Lakowicz, J. R., Black, L. W. 2010. DNA crunching by a viral packaging motor: Compression of a procapsid-portal stalled Y-DNA substrate. *Virology* 398, 224–232.

Renesto, P., Abergel, C., Decloquement, P., Moinier, D., Azza, S., Ogata, H., Fourquet, P., Gorvel, J.-P., Claverie, J.-M. 2006. Mimivirus giant particles incorporate a large fraction of anonymous and unique gene products. *J. Virology* 80(23): 11678–11685.

Sabanayagam, C. R., Oram, M., Lakowicz, J. R., Black, L. W. 2007. Viral DNA packaging studied by fluorescence correlation spectroscopy. *Biophys. J.* 93, L17–L19.

Schwartz, C., De Donatis, G. M., Zhang, H., Fang, H., Guo, P. 2013. Revolution rather than rotation of AAA + hexameric phi29 nanomotor for viral dsDNA packaging without coiling. *Virology* 443, 28–39.

Selvarajan, S. S., Zhao, H., Kamau, Y. N., Baines, J. D., Tang, L. 2013. The structure of the herpes simplex virus DNA-packaging terminase pUL15 nuclease domain suggests an evolutionary lineage among eukaryotic and prokary-otic viruses. *J. Virol.* 87, 7140–7148.

Serwer, P. 2010. A hypothesis for bacteriophage DNA packaging motors. *Viruses* 2, 1821–1843.

Sherratt, D. J., Arciszewska, L. K., Crozat, E., Graham, J. E., Grainge, I. 2010. The *Escherichia coli* DNA translocase FtsK. *Biochem. Soc. Trans.* 38, 395–398.

Shu, D., Zhang, H., Jin, J., Guo, P. 2007. Counting of six pRNAs of phi29 DNA-packaging motor with customized single molecule dual-view system. *EMBO J.* 26, 527–537.

Sivanathan, V., Emerson, J. E., Cornet, F., Sherratt, D. J., Arciszewska, L. K. 2009. KOPS-guided DNA translocation by FtsK safeguards Escherichia coli chro-mosome segregation. *Mol. Microbiol.* 71(4): 1031–1042.

Stevenson, E., Minton, N. P., Kuehne, S. A. 2015. The role of flagella in Clostridium difficile pathogenicity. *Trends Microbiol.* 23, 275–282.

Tang, J. H., Olson, N., Jardine, P. J., Girimes, S., Anderson, D. L., Baker, T. S. 2008. DNA poised for release in bacteriophage phi29. *Structure* 16, 935–943.

Thomas, D., Morgan, D., DeRosier, D. 1999. Rotational symmetry of the C ring and a mechanism for the flagellar rotary motor. *Proc. Natl. Acad. Sci. USA.* 96, 10134–10139.

Tyler, R. E., Ewing, S. G., Imperiale, M. J. 2007. Formation of a multiple protein complex on the adenovirus packaging sequence by the IVa2 protein. *J. Virol.* 81, 3447–3454.

Vale, R. 1993. Motor proteins. In: T. Kreis, R. Vale (Eds.), *Guidebook to the Cytoskeletal and Motor Proteins.* Oxford University Press, UK, pp. 175–211.

Van, G. N., Waksman, G., Remaut, H. 2011. Pili and flagella biology, structure, and biotechnological applications. *Prog. Mol. Biol. Transl. Sci.* 103, 21–72.

Walz, D., Caplan, S. 2000. An electrostatic mechanism closely reproducing observed behavior in the bacterial flagellar motor. *Biophys. J.* 78, 626–651.

Wang, M. D., Schnitzer, M. J., Yin, H., Landick, R., Gelles, J., Block, S. M. 1998. Force and velocity measured for single molecules of RNA polymerase. *Science* 282, 902–907.

Yogisharadhya, R., Bhanuprakash, V., Venkatesan, G., Balamurugan, V., Pandey, A. B., Shivachandra, S. B. 2012. Comparative sequence analysis of poxvirus A32 gene encoded ATPase protein and carboxyl terminal heterogeneity of Indian orf viruses. *Vet. Microbiol.* 156, 72–80.

Yoshida, M., Muneyuki, E., Hisabori, T. 2003. ATP synthase–A marvellous rotary engine of the cell. *Nat. Rev. Mol. Cell. Biol.* 2(9), 669–677.

Zauberman, N., Mutsafi, Y., Halevy, D. B., Shimoni, E., Klein, E., Xiao, C., Sun, S., Minsky, A. 2008. Distinct DNA exit and packaging portals in the virus Acanthamoeba polyphaga mimivirus. *PLoS Biol.* 6, e114.

Zhang, H., Endrizzi, J. A., Shu, Y., Haque, F., Sauter, C., Shlyakhtenko, L. S., Lyubchenko, Y., Guo, P., Chi, Y. I. 2013. Crystal structure of 3WJ core revealing divalent ion-promoted thermostability and assembly of the Phi29 hexameric motor pRNA. *RNA* 19, 1226–1237.

Zhang, H., Schwartz, C., De Donatis, G. M., Guo, P. 2012. "Push through one-way valve" mechanism of viral DNA packaging. *Adv. Virus Res.* 83, 415–465.

Zhao, Z., Khisamutdinov, E., Schwartz, C., Guo, P. 2013. Mechanism of one-way traffic of hexameric phi29 DNA packaging motor with four electropositive relaying layers facilitating anti-parallel revolution. *ACS Nano* 7, 4082–4092.

chapter two

Structure of revolving biomotors

2.1 Hexameric arrangement of motor components

In 1978, structural studies of viruses (Caspar and Klug, 1962; Carrascosa et al., 1982; Kochan et al., 1984; Bazinet and King, 1985; Valpuesta et al., 1992; Agirrezabala et al., 2005; Doan and Dokland, 2007; Cardarelli et al., 2010) led to the popular fivefold/sixfold symmetric mismatch gearing mechanism (Hendrix, 1978). Twenty years later, the pRNA complex on the phi29 bacteriophage was first shown to be hexameric by Guo et al. (Guo et al., 1998; Zhang et al., 1998) (published in *Cell* (Hendrix, 1998)). Despite some divergent models indicating this pRNA as a pentamer (Morais et al., 2008; Yu et al., 2010; Chistol et al., 2012), the hexameric configuration of pRNA was verified by various approaches, such as cryo-electron micros-copy (Cryo-EM) (Ibarra et al., 2000), biochemical analysis (Guo et al., 1998; Hendrix, 1998; Zhang et al., 1998), single-molecule photobleaching step counting (Shu et al., 2007), electron microscopy (EM) (Moll and Guo, 2007; Xiao et al., 2008), and crystallization studies (Zhang et al., 2013). The photo-bleaching study of pRNA was carried out under the active packaging of the phi29 motor, excluding the possibility that the pRNA complex functions as a pentamer with one of its subunit falling off after packaging initiation (Simpson et al., 2000; Morais et al., 2001, 2008). Members of the ASCE super-family mostly operate with a hexameric arrangement of its components (Chen et al., 2002; Iyer et al., 2004; Willows et al., 2004; Aker et al., 2007; Mueller-Cajar et al., 2011; Wang et al., 2011). Many revolving motors belong to the ASCE superfamily, including the FtsK hexamer and phi29 ATPase, which has been recently proven to adopt a hexamer configuration as its final oligomeric state (Schwartz et al., 2013) by virion assembly inhibition assays, binomial distribution analysis, qualitative DNA-binding assays, capillary electrophoresis (CE) assays, and *electrophoretic mobility shift assays* (EMSA) (Chen et al., 1997; Trottier and Guo, 1997; Schwartz et al., 2013) (Figure 2.1).

2.2 DsDNA translocation of bacteria phages

At the late stage in viral replication, the genome of linear dsDNA bacte-riophages and animal viruses translocate their genomic DNA into a pre-assembled protein shell, known as a procapsid. Due to the high compression of the DNA in the procapsid, this entropically unfavorable DNA packag-ing process is accomplished by a motor utilizing ATP (Guo et al., 2016).

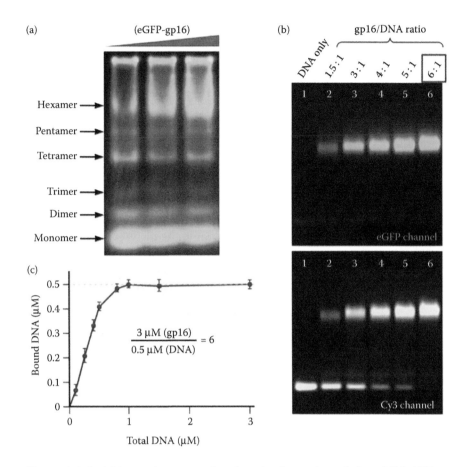

Figure 2.1 Stoichiometric assays showing the formation of the phi29 ATPase hexamer. (a) Native gel revealed six oligomeric states of the ATPase. (b) Slab gel showing the binding of ATPase to dsDNA in 6:1 ratio; imaged in GFP (upper) and Cy3 channels (lower) for ATPase and dsDNA, respectively. (c) Quantification by titration of ATPase and DNA. The concentration of bound DNA plateaus at a molar ratio of 6:1. (Adapted with permission from Schwartz, C. et al., 2013. *Virology* 443, 20–27.)

The motor involves three co-axial rings: a connector, a DNA packaging enzyme, and in phi29 a hexameric RNA ring. The connector is a dodecameric structure with a central channel that allows the genome to enter into the procapsid during genome packaging, and to exit during host infection. The DNA packaging enzyme is a ring consisting of six copies of a DNA-dependent ATPase (Guo et al., 1987). In phi29, the hexameric RNA ring serves as a linker between the hexameric ATPase ring and the motor channel (Guo et al., 2016).

2.3 DsDNA translocases of the FtsK/SpoIIIE superfamily

The FtsK motor contains three functional components: one for DNA translocation, one for orientation control, and one for anchoring itself to the bacterial membrane (Figure 2.2) (Demarre et al., 2013). The N-terminal domain of FtsK consists of four transmembrane helices and interacts with FtsZ, thus leading to its localization and to the forming septum at mid-cell just prior to division (Yu et al., 1998a; Dubarry et al., 2010). FtsK also interacts with a number of downstream cell division proteins to localize them to the divisome and allow cell division to progress (Dubarry et al., 2010). The N and C domains are connected by a linker domain rich in proline and glutamine, which also aids in interactions to localize the cell division apparatus. The C-terminus consists of the DNA translocation motor ATPase, which drives DNA away from the forming septum and also participates in XerCD site-specific recombination at *dif* sites to resolve chromosome dimers.

The C-terminal DNA translocation motor component of FtsK can be further subdivided into three domains, α, β, and γ (Massey et al., 2006). The α and β domains assemble into a hexameric ring-shaped channel, through

α   β   γ

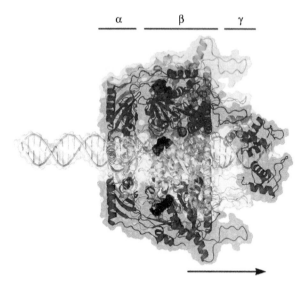

Figure 2.2 Illustration of FtsK motor loaded at a KOPS site. The N-termini of the γ domains are located on one side of the complex where they would connect to the motor domains of FtsK. This leads to loading of the motor to one side of the KOPS site so that the motor is pointing in a defined direction, giving the motor its subsequent directional translocation (black arrow denotes the direction the motor would move along the DNA). (Adapted with permission from Guo, P. et al. 2016. *Microbiol Mol Biol Rev.* 80(1), 161–186.)

which the dsDNA substrate is threaded (Massey et al., 2006), as revealed by EM. The rings formed by the α and β domains are separated by ~10-Å cleft and connected by two strands. The γ subdomain acts as both a protein–protein interaction domain and DNA-binding domain. It activates Xer-mediated recombination at *dif* (Grainge et al., 2011) and recognizes and binds to the specific 8-bp chromosomal sequences (GGGNAGGG) (Figures 2.2 and 2.3c) (Lowe et al., 2008) known as FtsK orienting polarized sequences (KOPS) (Bigot et al., 2005; Levy et al., 2005). The KOPS sequence acts as a recognition loading site for FtsK and "determines" the

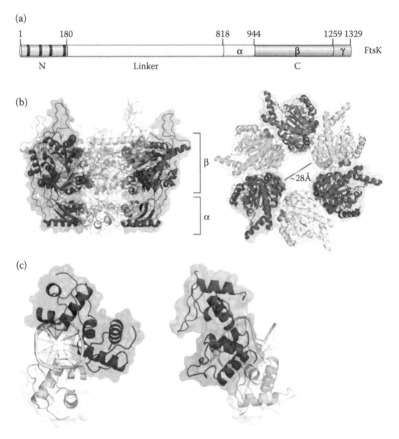

Figure 2.3 Structure of the FtsK motor protein. (a) Cartoon representation of the *E. coli* FtsK protein domain structure. The N-terminal domain is in red with each transmembrane domain represented by a black box. (b) Side view and top view (from the β domain side) of the hexameric FtsK motor protein structure (PDB: 2IUU). Bound nucleotide (ADP) is shown in black in a space-filling model. (c) Two views of the structure of three γ domains bound to a KOPS DNA. (Adapted with permission from Guo, P. et al. 2016. *Microbiol Mol Biol Rev.* 80(1), 161–186.)

directionality of translocation, which is toward the XerCD-*dif* site, which indicates that the directional translocation of DNA substrate is sequence-dependent. Furthermore, an active translocating FtsK appears to ignore further KOPS sequences and reads through them (Lowe et al., 2008; Lee et al., 2012).

Like FtsK, the conserved C-terminal domain of SpoIIIE also harbors three subdomains of α, β, and γ (Kaimer and Graumann, 2011; Fiche et al., 2013), which hexamerize and accommodate the dsDNA substrate in its central channel ring.

2.4   DNA translocases of NCLDV superfamily

NCLDV packaging ATPases and HerA has conserved C-terminal motor domain of FtsK (Iyer et al., 2004; Burrough et al., 2007). Repression of the putative FtsK-type packaging ATPase (gene *A32L*) in vaccinia virus resulted in the formation of DNA-deficient, non-infectious virus particles (Cassetti et al., 1998). Membrane tethering function in HerA is carried out by a separate small transmembrane protein called MJ1617 (Iyer et al., 2004). Although Mimivirus and other NCLDV packaging ATPases lack the obvious transmembrane region, they all show the presence of a pore lining region (Nugent, 2012).

References

Agirrezabala, X., Martin-Benito, J., Valle, M., Gonzalez, J. M., Valencia, A., Valpuesta, J. M., Carrascosa, J. L. 2005. Structure of the connector of bacteriophage T7 at 8A resolution: Structural homologies of a basic component of a DNA translocating machinery. *J. Mol. Biol.* 347, 895–902.

Aker, J., Hesselink, R., Engel, R., Karlova, R., Borst, J. W., Visser, A. J. W. G., de Vries, S. C. 2007. *In vivo* hexamerization and characterization of the *Arabidopsis* AAA ATPase CDC48A complex using forster resonance energy transfer-fluorescence lifetime imaging microscopy and fluorescence correlation spectroscopy. *Plant Physiol.* 145, 339–350.

Bazinet, C., King, J. 1985. The DNA translocation vertex of dsDNA bacteriophages. *Ann. Rev. Microbiol.* 39, 109–129.

Bigot, S., Saleh, O. A., Aravind, L., Pages, C., El, K. M., Dennis, C., Grigoriev, M., Allemand, J. F., Barre, F. X., Cornet, F. 2005. KOPS: DNA motifs that control *E. coli* chromosome segregation by orienting the FtsK translocase. *EMBO J.* 24, 3770–3780.

Burroughs, A., Iyer, L., Aravind, L. 2007. Comparative genomics and evolutionary trajectories of viral ATP dependent DNA-packaging systems. *Genome Dyn.* 3(R), 48.

Cardarelli, L., Lam, R., Tuite, A., Baker, L. A., Sadowski, P. D., Radford, D. R., Rubinstein, J. L. et al. 2010. The crystal structure of bacteriophage HK97 gp6: Defining a large family of head-tail connector proteins. *J. Mol. Biol.* 395, 754–768.

Carrascosa, J. L., Vinuela, E., Garcia, N., Santisteban, A. 1982. Structure of the head-tail connector of bacteriophage phi 29. *J. Mol. Biol.* 154, 311–324.

Caspar, D. L. D., Klug, A. 1962. Physical principles in the construction of regular viruses. *Cold Spring Harb. Symp. Quant. Biol.* 27, 1–24.

Cassetti, M. C., Merchlinsky, M., Wolffe, E. J., Weisberg, A. S., Moss, B. 1998. DNA packaging mutant: Repression of the vaccinia virus A32 gene results in noninfectious, DNA-deficient, spherical, enveloped particles. *J. Virol.* 72(7), 5769–5780.

Chen, C., Trottier, M., Guo, P. 1997. New approaches to stoichiometry determination and mechanism investigation on RNA involved in intermediate reactions. *Nucleic Acids Symp. Ser.* 36, 190–193.

Chen, Y. J., Yu, X., Egelman, E. H. 2002. The hexameric ring structure of the *Escherichia coli* RuvB branch migration protein. *J. Mol. Biol.* 319, 587–591.

Chistol, G., Liu, S., Hetherington, C. L., Moffitt, J. R., Grimes, S., Jardine, P. J., Bustamante, C. 2012. High degree of coordination and division of labor among subunits in a homomeric ring ATPase. *Cell* 151, 1017–1028.

Demarre, G., Galli, E., Barre, F. X. 2013. The FtsK family of DNA pumps. *Adv. Exp. Med. Biol.* 767, 245–262.

Doan, D. N., Dokland, T. 2007. The gpQ portal protein of bacteriophage P2 forms dodecameric connectors in crystals. *J. Struct. Biol.* 157, 432–436.

Dubarry, N. et al. 2010. Multiple regions along the Escherichia coli FtsK protein are implicated in cell division. *Mol Microbiol.* 78(5), 1088–1100.

Fiche, J. B., Cattoni, D. I., Diekmann, N., Langerak, J. M., Clerte, C., Royer, C. A., Margeat, E., Doan, T., Nollmann, M. 2013. Recruitment, assembly, and molecular architecture of the SpoIIIE DNA pump revealed by superresolution microscopy. *PLoS Biol.* 11, e1001557.

Grainge, I. et al. 2011. Activation of XerCD-dif recombination by the FtsK DNA translocase. *Nucleic Acids Res.* 39(12), 5140–5148.

Guo, P., Noji, H., Yengo, C. M., Zhao, Z., Grainge, I. 2016. Biological nanomotors with a revolution, linear, or rotation motion mechanism. *Microbiol Mol Biol Rev.* 80(1), 161–186.

Guo, P., Peterson, C., Anderson, D. 1987. Initiation events in in vitro packaging of bacteriophage phi29 DNA-gp3. *J. Mol. Biol.* 197, 219–228.

Guo, P., Zhang, C., Chen, C., Trottier, M., Garver, K. 1998. Inter-RNA interaction of phage phi29 pRNA to form a hexameric complex for viral DNA transportation. *Mol. Cell* 2, 149–155.

Hendrix, R. W. 1978. Symmetry mismatch and DNA packaging in large bacteriophages. *Proc. Natl. Acad. Sci. USA* 75, 4779–4783.

Hendrix, R. W. 1998. Bacteriophage DNA packaging: RNA gears in a DNA transport machine (Minireview). *Cell* 94, 147–150.

Ibarra, B., Caston, J. R., Llorca, O., Valle, M., Valpuesta, J. M., Carrascosa, J. L. 2000. Topology of the components of the DNA packaging machinery in the phage phi29 prohead. *J. Mol. Biol.* 298, 807–815.

Iyer, L. M., Leipe, D. D., Koonin, E. V., Aravind, L. 2004. Evolutionary history and higher order classification of AAA plus ATPases. *J. Struct. Biol.* 146, 11–31.

Kaimer, C., Graumann, P. L. 2011. Players between the worlds: Multifunctional DNA translocases. *Curr. Opin. Microbiol.* 14, 719–725.

Kochan, J., Carrascosa, J. L., Murialdo, H. 1984. Bacteriophage lambda preconnectors: Purification and structure. *J. Mol. Biol.* 174, 433–447.

Lee, J. Y. et al. 2012. Single-molecule imaging of DNA curtains reveals mechanisms of KOPS sequence targeting by the DNA translocase FtsK. *Proc Natl Acad Sci U S A.* 109(17), 6531–6536.

Levy, O., Ptacin, J. L., Pease, P. J., Gore, J., Eisen, M. B., Bustamante, C., Cozzarelli, N. R. 2005. Identification of oligonucleotide sequences that direct the movement of the *Escherichia coli* FtsK translocase. *Proc. Natl. Acad. Sci. USA* 102, 17618–17623.

Lowe, J. et al. 2008. Molecular mechanism of sequence-directed DNA loading and translocation by FtsK. *Mol Cell.* 31(4), 498–509.

Massey, T. H. et al. 2006. Double-stranded DNA translocation: Structure and mechanism of hexameric FtsK. *Mol Cell.* 23(4), 457–469.

Moll, D., Guo, P. 2007. Grouping of ferritin and gold nanoparticles conjugated to pRNA of the phage phi29 DNA-packaging motor. *J. Nanosci. Nanotech.* 7, 3257–3267.

Morais, M. C., Koti, J. S., Bowman, V. D., Reyes-Aldrete, E., Anderson, D., Rossman, M. G. 2008. Defining molecular and domain boundaries in the bacteriophage phi29 DNA packaging motor. *Structure* 16, 1267–1274.

Morais, M. C., Tao, Y., Olsen, N. H., Grimes, S., Jardine, P. J., Anderson, D., Baker, T. S., Rossmann, M. G. 2001. Cryoelectron-microscopy image reconstruction of symmetry mismatches in bacteriophage phi29. *J. Struct. Biol.* 135, 38–46.

Mueller-Cajar, O., Stotz, M., Wendler, P., Hartl, F. U., Bracher, A., Hayer-Hartl, M. 2011. Structure and function of the AAA+ protein CbbX, a red-type Rubisco activase. *Nature* 479, 194–199.

Nugent, T., Jones, D. T. 2012. Detecting pore-lining regions in transmembrane protein sequences. *BMC Bioinformatics* 13(1), 169.

Schwartz, C., De Donatis, G. M., Fang, H., Guo, P. 2013. The ATPase of the phi29 DNA-packaging motor is a member of the hexameric AAA+ superfamily. *Virology* 443, 20–27.

Shu, D., Zhang, H., Jin, J., Guo, P. 2007. Counting of six pRNAs of phi29 DNA-packaging motor with customized single molecule dual-view system. *EMBO J.* 26, 527–537.

Simpson, A. A., Tao, Y., Leiman, P. G., Badasso, M. O., He, Y., Jardine, P. J., Olson, N. H. et al. 2000. Structure of the bacteriophage phi29 DNA packaging motor. *Nature* 408, 745–750.

Trottier, M., Guo, P. 1997. Approaches to determine stoichiometry of viral assembly components. *J. Virol.* 71, 487–494.

Valpuesta, J. M., Fujisawa, H., Marco, S., Carazo, J. M., Carrascosa, J. 1992. Three-dimensional structure of T3 connector purified from overexpressing bacteria. *J. Mol. Biol.* 224, 103–112.

Wang, F., Mei, Z., Qi, Y., Yan, C., Hu, Q., Wang, J., Shi, Y. 2011. Structure and mechanism of the hexameric MecA-ClpC molecular machine. *Nature* 471, 331–335.

Willows, R. D., Hansson, A., Birch, D., Al-Karadaghi, S., Hansson, M. 2004. EM single particle analysis of the ATP-dependent BchI complex of magnesium chelatase: An AAA(+) hexamer. *J. Struct. Biol.* 146, 227–233.

Xiao, F., Zhang, H., Guo, P., 2008. Novel mechanism of hexamer ring assembly in protein/RNA interactions revealed by single molecule imaging. *Nucleic Acids Res.* 36, 6620–6632.

Yu, J., Moffitt, J., Hetherington, C. L., Bustamante, C., Oster, G. 2010. Mechanochemistry of a viral DNA packaging motor. *J. Mol. Biol.* 400, 186–203.

Yu, X. C. et al. 1998. Localization of cell division protein FtsK to the Escherichia coli septum and identification of a potential N-terminal targeting domain. *J Bacteriol.* 180(5), 1296–1304.

Zhang, F., Lemieux, S., Wu, X., St.-Arnaud, S., McMurray, C. T., Major, F., Anderson, D. 1998. Function of hexameric RNA in packaging of bacteriophage phi29 DNA *in vitro*. *Mol. Cell* 2, 141–147.

Zhang, H., Endrizzi, J. A., Shu, Y., Haque, F., Sauter, C., Shlyakhtenko, L. S., Lyubchenko, Y., Guo, P., Chi, Y. I. 2013. Crystal structure of 3WJ core revealing divalent ion-promoted thermostability and assembly of the phi29 hexameric motor pRNA. *RNA* 19, 1226–1237.

chapter three

Structure of rotation motors

3.1   Structure of flagellar motors

Flagellar motors are composed of two parts: the rotating part, or rotor, and the nonrotating part, or stator (basal body). The rotor is made of four rings (C, coaxial MS, L, and P) and the rod. It is connected to the filament and hook (tubular structure at the base region), and anchored in the cytoplasmic membrane and cell wall (Terashima et al., 2008; Thormann and Paulick, 2010). The stator is a membrane-embedded energy converter composed of MotA/MotB complex with around 12–14 subunits that are organized around two rings in the cytoplasmic membrane. Ion flow through stator (MotAB) is believed to be coupled with force generation, generated by the interaction between rotor and stator components (Terashima et al., 2008; Thormann and Paulick, 2010). Rotation of the motor can be in both directions, clockwise and counterclockwise, and the direction can be switched spontaneously with the complex arrangement of the rotor.

3.2   Structure of F_oF_1 ATPase

F_oF_1 ATPase, located in the mitochondrial, bacterial, and chloroplast membranes, consists of two rotors, motors termed F_1 and F_o. They are located on opposite ends of the stator, a drive shaft known as γ, which connects the motors. Since they are connected, they work as a unit: one motor can cause the other to turn. As both motors share the γ shaft, they compete for control over it. When a steep proton flow moves through the F_o motor, it can force the F_1 motor to rotate by turning the γ axle and creating torque, forcing it to act as a generator and thus to produce ATP. This motor is composed of five subunits known as α, β, γ, δ, and ε. At least three α and three β, which are arranged in a hexameric ring around the γ subunit, are required to sustain ATP synthesis. F_o has at least three subunits, polypeptides referred to as A, B, and C. Three ring-shaped C subunits form its rotor portion.

F_1 and F_o are distinct in structure and function. F_1 can be reversibly dissociated from F_o, and shows strong ATP hydrolysis activity that is coupled with inner subunit rotation. Thereby, F_1 is termed F_1-ATPase. F_1 consists of $\alpha_3\beta_3\gamma_1\delta_1\varepsilon_1$ subunits in bacterial systems, and its minimum motor complex is the $\alpha_3\beta_3\gamma$ subcomplex (Figure 1.4). The catalytic reaction centers are located at the α–β interface, mainly on the β subunit. Crystal structure revealed that three α and three β subunits form the hexameric stator ring with a large central cavity, which accommodates half of the long coiled-coil structure of the rotary γ subunit (Figure 1.4) (Abrahams et al., 1994). The reported structure of the F_1-$F_o c$ complex of yeast ATP synthase showed that the other half of the coiled-coil of γ extends to bind to the rotor part of F_o (the oligomer ring of the c subunits) (Stock et al., 1999). The ε subunit binds to the side surface of the protruding part of the γ and also has close contact with $F_o c$. The δ subunit is located on the bottom tip of the $\alpha_3\beta_3$ ring. The rotation of γ is the result of cooperative conformational change in the β subunits, exerting the catalysis (Uchihashi et al., 2011). These catalytic sites can adopt different binding states. The first binds with Mg-AMP-PNP (adenosine 5'-(β,γ-imino)-triphosphate), the second with Mg-ADP, and the third is empty; they are referred to as β_{ATP}, β_{ADP}, and β_{empty}, respectively. The two ligand-bound states of β (β_{ATP} and β_{ADP}) show closed conformation, swinging the upwardly protruding C-terminal domain toward the central γ subunit (Figure 1.4b). On the other hand, β_{empty} assumes the open conformation with the C-terminal domain opening outwardly.

F_o is composed of $a_1 b_1 c_{8-15}$ subunits (Figure 3.1). The number of the c subunit differs among species while the model bacterial F_o has 10 c subunits. F_o generates torque coupled with proton translocation across a membrane down pmf. Compared with F_1, F_o is less studied. This is mainly because F_o is a membrane-embedded complex and a highly hydrophobic protein complex, causing difficulties in purification, handling, and activity measurement. The atomic structure of the whole complex of F_o is still not available, although the isolated oligomer ring of the c subunit was crystalized for several species (Stock et al., 1999; Meier et al., 2005; Pogoryelov et al., 2009; Vollmar et al., 2009; Watt et al., 2010; Preiss et al., 2013) (Figure 3.1a). While F_1 has a pseudocircular symmetry in the stator ring, F_o has a circular symmetry in the rotor part; the c subunits form the oligomeric ring that rotates against the stator $a_1 b_2$ complex of F_o, which forms the stator complex with the $\alpha_3\beta_3$ ring in the whole complex of ATP synthase. The proton translocation channel is formed by the a subunit; and the c oligomeric ring, that is, protons, is translocated between the rotor and stator parts. It is thought that upon each proton translocation, the c ring rotates against the a subunit by the angle for a single c subunit in the c ring; 36° steps for the c_{10} ring.

In the whole complex of $F_o F_1$ ATP synthase, the c ring binds to the $\gamma\varepsilon$ complex to form the common rotary shaft of ATP synthase (Stock et al., 1999). The b dimer complex spans the lipid membrane extending toward the headpiece of F_1 to bind to the δ subunit that sits on the bottom hollow of the $\alpha_3\beta_3$ ring (Walker and Dickson, 2006). The b_2-δ complex forms the peripheral stalk to hold the stator parts of F_1 and F_o (the

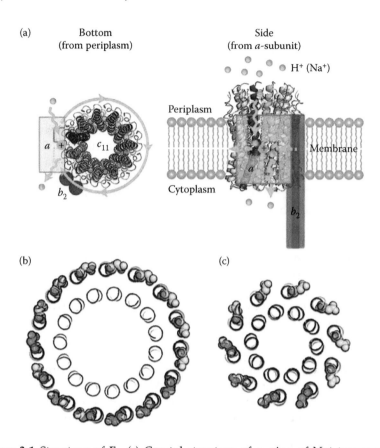

Figure 3.1 Structure of F_o. (a) Crystal structure of c_{11}-ring of Na⁺-transporting F_o from *Ilyobacter tartaricus* (PDB code; 1YCE). The blue spheres in the middle of the c_{11}-ring represent bound Na⁺ ions. Proton transferring between the *a* and *c* subunits accompanies the rotation of the *c*-ring. Two *c*-subunit monomers at the interface of the *a* subunit are shown in red and green, respectively. (b) "Ion-locked" conformation of cGlu62 (yellow sphere representation) in the crystal structure of H⁺-transporting c_{15}-ring from *Spirulina platensis* (PDB code; 2WIE). (c) "Ion-unlocked" conformation of cGlu59 (yellow sphere representation) in the crystal structure of H⁺-transporting c_{10}-ring from yeast mitochondria (PDB code; 3U2F). (Adapted with permission from Guo, P. et al. 2016. *Microbiol. Mol. Biol. Rev.* 80(1), 161–186.)

$\alpha_3\beta_3$ ring and the *a* subunit) stably to transmit the torque efficiently between each other. To achieve the energy interconversion between pmf and free energy of ATP hydrolysis, ΔG_{ATP}, the rotation directions of F_o are opposite to that of F_1. The rotation direction of F_oF_1 ATP synthase is determined by the magnitude relation of driving forces of F_o and F_1, each derived from pmf and ΔG_{ATP}. The driving force is defined as the number of driving reactions per rotation divided by 2π. Therefore, when $n \cdot \text{pmf} > -3 \cdot \Delta G_{ATP}$, where n represents the number of c subunits in the c ring of F_o, F_o governs the rotation, reversibly rotating the common rotary shaft (the $\gamma\varepsilon$-c ring complex) against F_1 that induces the reverse reaction of ATP hydrolysis: ATP synthesis. When $n \cdot \text{pmf} < -3 \cdot \Delta G_{ATP}$, F_1 rotates the rotary shaft, enforcing F_o to pump protons to build pmf. Thus, n is physiologically critical to determine the chemical equilibrium of ATP synthesis/hydrolysis. It is thought that n has been tuned under evolutional pressure to meet the physiological requirements (von Ballmoos et al., 2009). Actually, F_o has only eight copies of the c subunit in mammalian mitochondria where respiratory chains produce large pmf (Watt et al., 2010). On the other hand, in some bacteria and chloroplast with supposedly low pmf, n is large; 13–15 (Pogoryelov et al., 2009; Vollmar et al., 2009; Preiss et al., 2013, 2014).

From a mechanistic point of view, n would be important for smooth energy transmission. In the majority of species studied so far, n is a noninteger multiple of 3 such as 8, 10, and 11. To date, a 9-mer ring has not been found. One rare exception is the 15-mer ring of F_o from *Spirulina platensis* (Pogoryelov et al., 2009). A rotor ring with threefold symmetry was also found in a bacterial type of vacuole type H^+-ATPase (Toei et al., 2007) that shares remarkably similar structure and working mechanism, although there are some significant differences. Thus, a noninteger multiple of 3 is not an obligation for rotary ATPases. However, the apparent preference of non-threefold symmetry in the proton-conducting unit would represent the intrinsic nature of the rotary coupling mechanism of ATP synthase. One intriguing and reasonable explanation for this is that it would provide a relatively smooth rotary potential surface for the rotor complex, allowing the $\gamma\varepsilon$-c ring complex to avoid being trapped in a deep potential minimum where the potential minimums of F_1 and F_o overlap.

References

Abrahams, J. P., Leslie, A. G., Lutter, R., Walker, J. E. 1994. Structure at 2.8 A resolution of F1-ATPase from bovine heart mitochondria. *Nature* 370, 621–628.

Guo, P., Noji, H., Yengo, C. M., Zhao, Z., Grainge, I. 2016. Biological nanomotors with a revolution, linear, or rotation motion mechanism. *Microbiol. Mol. Biol. Rev.* 80(1), 161–186.

Meier, T., Polzer, P., Diederichs, K., Welte, W., Dimroth, P. 2005. Structure of the rotor ring of F-type Na+-ATPase from *Ilyobacter tartaricus*. *Science* 308, 659–662.

Pogoryelov, D., Yildiz, O., Faraldo-Gomez, J. D., Meier, T. 2009. High-resolution structure of the rotor ring of a proton-dependent ATP synthase. *Nat. Struct. Mol. Biol.* 16, 1068–1073.

Preiss, L., Klyszejko, A. L., Hicks, D. B., Liu, J., Fackelmayer, O. J., Yildiz, O., Krulwich, T. A., Meier, T. 2013. The c-ring stoichiometry of ATP synthase is adapted to cell physiological requirements of alkaliphilic *Bacillus pseudofirmus* OF4. *Proc. Natl. Acad. Sci. USA* 110, 7874–7879.

Preiss, L., Langer, J. D., Hicks, D. B., Liu, J., Yildiz, O., Krulwich, T. A., Meier, T. 2014. The c-ring ion binding site of the ATP synthase from *Bacillus pseudofirmus* OF4 is adapted to alkaliphilic lifestyle. *Mol. Microbiol.* 92, 973–984.

Stock, D., Leslie, A. G., Walker, J. E. 1999. Molecular architecture of the rotary motor in ATP synthase. *Science* 286, 1700–1705.

Terashima, H., Kojima, S., Homma, M. 2008. Flagellar motility in bacteria structure and function of flagellar motor. *Int. Rev. Cell Mol. Biol.* 270, 39–85.

Thormann, K. M., Paulick, A. 2010. Tuning the flagellar motor. *Microbiology* 156(Pt 5), 1275–1283.

Toei, M., Gerle, C., Nakano, M., Tani, K., Gyobu, N., Tamakoshi, M., Sone, N. et al. 2007. Dodecamer rotor ring defines H+/ATP ratio for ATP synthesis of prokaryotic V-ATPase from *Thermus thermophilus*. *Proc. Natl. Acad. Sci. USA* 104, 20256–20261.

Uchihashi, T., Iino, R., Ando, T., Noji, H. 2011. High-speed atomic force microscopy reveals rotary catalysis of rotorless F-1-ATPase. *Science* 333, 755–758.

Vollmar, M., Schlieper, D., Winn, M., Buchner, C., Groth, G. 2009. Structure of the c(14) rotor ring of the proton translocating chloroplast ATP synthase. *J. Biol. Chem.* 284, 18228–18235.

von Ballmoos, C., Wiedenmann, A., Dimroth, P. 2009. Essentials for ATP synthesis by F1F0 ATP synthases. *Annu. Rev. Biochem.* 78, 649–672.

Walker, J. E., Dickson, V. K. 2006. The peripheral stalk of the mitochondrial ATP synthase. *Biochim. Biophys. Acta.* 1757, 286–296.

Watt, I. N., Montgomery, M. G., Runswick, M. J., Leslie, A. G. W., Walker, J. E. 2010. Bioenergetic cost of making an adenosine triphosphate molecule in animal mitochondria. *Proc. Natl. Acad. Sci. USA* 107, 16823–16827.

chapter four

Structure of linear motors

4.1   Overall design of linear motors

Myosin, kinesin, and dynein power many different essential cellular functions. The mechanical work produced by myosin drives many biological processes including muscle contraction, movement of cargo or organelles on actin filaments, membrane tension generation, endocytosis, and exocytosis (Hartman & Spudich, 2012). Myosin motors can also participate in signal transduction and transcription (Quintero et al., 2010; Hardie et al., 2012; Zorca et al., 2015). Kinesin and dynein carry out much of the long distance transport in neurons and dynein powers the bending of cilia and flagella such as the bending of cilia and flagella (Hirokawa & Takemura, 2005; Lindemann & Lesich, 2010).

Most linear motors possess head, neck, and tail domains, with the head allowing and directing movement, the tail identifying its cargo, and the neck connecting the head and tail. Unlike the other two types of motors, they contain all the elements capable of converting chemical energy into mechanical work. The majority of these motors have two heads that each can bind and hydrolyze ATP. The two heads also allow the motor to move along filament tracks and determine the direction of movement. The tail domain allows incorporation into filaments or determines the identity of the motor's cargo and thus its biological function. Most linear motors consist of a heavy chain containing the motor, neck, and tail and associated light chains that serve to stabilize/regulate the motor and/or allow protein-protein interactions. Kinesin and myosin share a similar structural fold and are members of the P-loop NTPases that contain a conserved nucleotide binding region (Kull et al., 1998). Globally, their motor domains have a similar architecture with a highly conserved nucleotide-binding region communicating with both the effector binding (actin or microtubule) and the force-generating regions (Kull & Endow, 2013). Dynein is a more complex ring-shaped motor that has six ATP binding regions and is a member of the AAA+ family. Only four of the six nucleotide binding sites bind nucleotide while a single site controls the movement (Bhabha et al., 2016; Roberts et al., 2013).

4.2 Structure of myosins

The structure of the myosin motor domain (Figure 4.1) is highly conserved among various classes of myosins (Geeves & Holmes, 1999; Vale & Milligan, 2000). Therefore, the conformational pathway of the myosin ATPase cycle is hypothesized to be similar in all myosin motors while variation in the kinetic and equilibrium constants of the conformational changes allows for myosins to be fine-tuned for performing specific cellular functions (De La Cruz & Ostap, 2004; Heissler & Sellers, 2016). The motor domain contains ATP binding (Nucleotide Binding Pocket—NBP) and actin binding (Actin Binding Cleft-Cleft) motifs, which are coupled with the reciprocal movement of the lever arm region during the recovery and power stroke states of the ATPase cycle (Sweeney & Houdusse, 2010). However, the mechanism of allosteric communication between different sub-domains of the motor remains a crucial question in the field today. Any perturbation to these communication pathways can disrupt the motor properties and lead to diseased states (Petit & Richardson, 2009; Williams & Lopes, 2011; Spudich et al., 2016). While the allosteric communication between the nucleotide and actin binding regions has been extensively studied (Conibear et al., 2003; Malnasi-Csizmadia et al., 2005; Kintses et al., 2007), there remains outstanding questions regarding the coupling of actin binding, product release, and the position of the lever arm (Malnasi-Csizmadia & Kovacs, 2010). A long-standing question in the actomyosin field is the precise timing of force generation and its relationship to the kinetics of lever-arm swing.

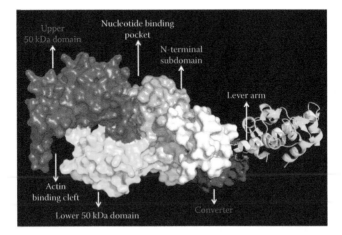

Figure 4.1 Crystal structure showing the subdomains of myosin along with the actin, nucleotide binding, and lever arm regions (PDB code: 1W7J). (Adapted from Trivedi, D. V. 2014. Allosteric communication and force generation in myosin motors. Doctoral dissertation. Pennsylvania State University, University Park, PA; Guo, P. et al., 2016. *Microbiol. Mol. Biol. Rev.* 80, 161–186, with permission.)

4.2.1 Nucleotide binding region

The nucleotide binding pocket of myosin coordinates nucleotide binding, cleavage of the phosphoanhydride bond of ATP, and the sequential release of products that is crucial for converting chemical energy into mechanical work (Lymn & Taylor, 1971; Eisenberg & Greene, 1980; De La Cruz & Ostap, 2004). The ATP molecule is coordinated in the NBP by three highly conserved structural elements, switch I, switch II, and the P-loop (Kull et al., 1998; Vale & Milligan, 2000) (Figure 4.2). The family of P-loop NTPases, including G-proteins, kinesins, and myosins, is thought to have evolved from a common ancestor (Kull et al., 1998). Switch I has been reported to be an important element that coordinates the sequential release of products and transmits information from the NBP to the actin binding cleft in myosins (Kintses et al., 2007). Switch II is another well-conserved element that forms a salt bridge with Switch I and interacts

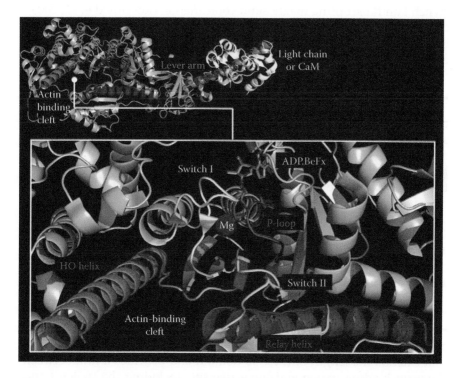

Figure 4.2 Crystal structure of myosin showing the key structural elements involved in the coordination of ATP and energy transduction mechanism (PDB code: 1W7J). (Adapted from Trivedi, D. V. 2014. Allosteric communication and force generation in myosin motors. Doctoral dissertation. Pennsylvania State University, University Park, PA; Guo, P. et al., 2016. *Microbiol. Mol. Biol. Rev.* 80, 161–186, with permission.)

with the γ-phosphate of ATP, which is essential for catalysis (Sasaki et al., 1998; Onishi et al., 2002). Coordination of the α and β phosphates of ATP is accomplished by the P loop. The coordination of magnesium (Mg) is through the oxygen on the β and γ phosphates and several residues in the switch I region (Rayment et al., 1996; Swenson et al., 2014). The γ-phosphate of ATP plays a central role in the interaction of the switch elements and the P-loop with ATP. This explains why ATP, and not ADP, can lead to the weak actin-binding state of myosin and also induces the recovery stroke of the lever arm. The binding of ATP to the active site results in the switch elements adopting a "closed" conformation, which leads to an opening of the actin binding cleft (Coureux et al., 2004). The closing of the switch elements also gets communicated to the lever-arm region via the relay helix, which results in formation of the pre-powerstroke state (Fischer et al., 2005; Malnasi-Csizmadia et al., 2005; Nesmelov et al., 2011) (Figure 4.2). Actin binding can stimulate the release of ADP by changing the conformation of switch I and disrupting magnesium coordination (Rosenfeld et al., 2005). It has been speculated that the coupling associated with actin and ADP binding requires the coordination of bound magnesium (Hannemann et al., 2005; Rosenfeld et al., 2005; Swenson et al., 2014). Moreover, two different states of switch I have been reported when MgADP is bound in the pocket, and both of these states bind ADP differently (Hannemann et al., 2005). In rapidly contracting muscle fibers, ADP release has been shown to be a major determinant of the maximum shortening velocity and has been speculated to be a central step for sensing load on the myosin crossbridge, thus making it a strain-sensitive step (Nyitrai & Geeves, 2004). Since the lever arm senses the load, there must be allosteric communication between the lever arm and the NBP to modulate the load-dependent release of ADP.

4.2.2 *Actin binding region*

The actin binding region contains a large cleft that separates the 50 kDa domain into upper (U50) and lower (L50) sub-domains (Rayment et al., 1993a, b; Houdusse & Sweeney, 2001). The open conformation of the cleft is associated with weak actin binding while strong binding occurs in the closed cleft conformation (Coureux et al., 2004). It is proposed that the open–closed conformation of the actin binding cleft is coupled to the closed–open conformation of switch I. There is a long alpha helix (HO) that extends the length of the U50 domain and provides a link between the actin binding region and the active site, including the switch I and switch II regions (Ovchinnikov et al., 2010). Conformational changes in the HO helix were demonstrated by monitoring an intrinsic tryptophan residue during the ATP and actin binding steps (Yengo et al., 2002a). The lower 50 kDa domain is connected to the relay-helix, which allows communication of the actin binding and lever arm regions (Agafonov et al.,

2009; Nesmelov et al., 2011; Muretta et al., 2013). The relay-helix transmits structural changes by changing from a kinked to a straight conformation during formation of the pre-powerstroke state and vice versa during the power stroke (Agafonov et al., 2009; Malnasi-Csizmadia et al., 2005; Nesmelov et al., 2011; Muretta et al., 2013). Thus, these two long alpha helices (relay and HO) that connect the actin binding, active site, and lever arm regions play crucial roles in allowing allosteric coupling between important regions of the myosin motor (Guo et al., 2016).

4.2.3 Lever arm region

The light chain binding region or neck of myosins is often referred to as the lever arm region since it is proposed to serve as the force generating element. A number of structural studies have demonstrated that the lever arm undergoes a large rotation in different nucleotide states (Irving et al., 1995; Forkey et al., 2003; Muretta et al., 2015; Trivedi et al., 2015). The lever arm consists of a long alpha helix that extends out from the globular motor domain and is stabilized by calmodulin or calmodulin-like light chains. Myosin V (MV) has served as a model for examining the role of the lever arm in actomyosin-based force generation. It is an unconventional, dimeric, and highly processive myosin involved in transporting vesicles and cargoes along actin filaments in cells (Mehta et al., 1999). It can walk along actin filaments using a hand-over-hand mechanism, taking 36 nm steps to transport an associated cargo (Yildiz et al., 2003), which is similar to the kinesin walking mechanism. The neck region of MV has 6 IQ-domains that can individually bind Calmodulin (CaM) and mechanically stiffen the lever arm to stabilize the position of the motor domain for coordinated stepping. The number of light chain binding sites varies in different myosins, which has allowed a direct test of the lever arm hypothesis that the length of the lever arm controls the distance a single myosin head can displace an actin filament per ATP hydrolyzed (Sakamoto et al., 2003; Moore et al., 2004). MV has served as an outstanding model to study actin-induced structural changes as its affinity for actin is much higher in the weak-binding states compared to conventional myosin II (Yengo et al., 2002b). Thus, using MV as a model system, structural differences between the weak and strong actin binding states can be elucidated with spectroscopic, kinetic, and structural studies (Volkmann et al., 2005; Sellers & Veigel, 2006; Sun et al., 2008; Jacobs et al., 2011) at lower actin concentrations. In addition, the crystal structure of MV has been solved in different nucleotide states using nucleotide analogs and by introducing mutations at key sites to stabilize structural states (Coureux et al., 2004; Llinas et al., 2015). The structure of MV bound to actin using cryo-EM has revealed key structural changes in the actin bound states of the motor (Volkmann et al., 2005; Wulf et al., 2016). Overall, the MV structural studies have revealed a framework to understand the structural elements of the conserved myosin ATPase cycle, while spectroscopic and biophysical studies have provided novel insights into the fundamental question of how the lever arm functions as a force generating element and allows the conversion of chemical energy into mechanical work.

References

Agafonov, R. V., Negrashov, I. V., Tkachev, Y. V., Blakely, S. E., Titus, M. A., Thomas, D. D., Nesmelov, Y. E. 2009. Structural dynamics of the myosin relay helix by time-resolved EPR and FRET. *Proc. Natl. Acad. Sci. USA* 106, 21625–21630.

Bhabha, G., Johnson, G. T., Schroeder, C. M., Vale, R. D. 2016. How dyein moves along microtubules. *Trends Biochem. Sci.* 41, 94–105.

Conibear, P. B., Bagshaw, C. R., Fajer, P. G., Kovacs, M., Malnasi-Csizmadia, A. 2003. Myosin cleft movement and its coupling to actomyosin dissociation. *Nat. Struct. Biol.* 10, 831–835.

Coureux, P. D., Sweeney, H. L., Houdusse, A. 2004. Three myosin V structures delineate essential features of chemo-mechanical transduction. *EMBO J.* 23, 4527–4537.

De la Cruz, E. M., Ostap, E. M. 2004. Relating biochemistry and function in the myosin superfamily. *Curr. Opin. Cell. Biol.* 16, 61–67.

Eisenberg, E., Greene, L. E. 1980. The relation of muscle biochemistry to muscle physiology. *Annu. Rev. Physiol.* 42, 293–309.

Fischer, S., Windshugel, B., Horak, D., Holmes, K. C., Smith, J. C. 2005. Structural mechanism of the recovery stroke in the myosin molecular motor. *Proc. Natl. Acad. Sci. USA* 102, 6873–6878.

Forkey, J. N., Quinlan, M. E., Shaw, M. A., Corrie, J. E., Goldman, Y. E. 2003. Three-dimensional structural dynamics of myosin V by single-molecule fluorescence polarization. *Nature* 422, 399–404.

Geeves, M. A., Holmes, K. C. 1999. Structural mechanism of muscle contraction. *Annu. Rev. Biochem.* 68, 687–728.

Guo, P., Noji, H., Yengo, C. M., Zhao, Z., Grainge, I. 2016. Biological nanomotors with revolution, linear, or rotation motion mechanism. *Microbiol. Mol. Biol. Rev.* 80, 161–186.

Hannemann, D. E., Cao, W., Olivares, A. O., Robblee, J. P., De La Cruz, E. M. 2005. Magnesium, ADP, and actin binding linkage of myosin V: Evidence for multiple myosin V-ADP and actomyosin V-ADP states. *Biochemistry* 44, 8826–8840.

Hardie, R. C., Satoh, A. K., Liu, C. H. 2012. Regulation of arrestin translocation by Ca2+ and myosin III in Drosophila photoreceptors. *J. Neurosci.* 32, 9205–9216.

Hartman, M. A., Spudich, J. A. 2012. The myosin superfamily at a glance. *J. Cell. Sci.* 125, 1627–1632.

Heissler, S. M., Sellers, J. R. 2016. Kinetic adaptations of myosins for their diverse cellular functions. *Traffic* 17, 839–859.

Hirokawa, N., Takemura, R. 2005. Molecular motors and mechanisms of directional transport in neurons. *Nat. Rev. Neurosci.* 6, 201–214.

Houdusse, A., Sweeney, H. L. 2001. Myosin motors: Missing structures and hidden springs. *Curr. Opin. Struct. Biol.* 11, 182–194.

Irving, M., St Claire Allen, T., Sabido-David, C., Craik, J. S., Brandmeier, B., Kendrick-Jones, J., Corrie, J. E., Trentham, D. R., Goldman, Y. E. 1995. Tilting of the light-chain region of myosin during step length changes and active force generation in skeletal muscle. *Nature* 375, 688–691.

Jacobs, D. J., Trivedi, D., David, C., Yengo, C. M. 2011. Kinetics and thermodynamics of the rate-limiting conformational change in the actomyosin V mechanochemical cycle. *J. Mol. Biol.* 407, 716–730.

Kintses, B., Gyimesi, M., Pearson, D. S., Geeves, M. A., Zeng, W., Bagshaw, C. R., Malnasi-Csizmadia, A. 2007. Reversible movement of switch 1 loop of myosin determines actin interaction. *EMBO J.* 26, 265–274.

Kull, F. J., Endow, S. A. 2013. Force generation by kinesin and myosin cytoskeletal motor proteins. *J. Cell. Sci.* 126, 9–19.

Kull, F. J., Vale, R. D., Fletterick, R. J. 1998. The case for a common ancestor: Kinesin and myosin motor proteins and G proteins. *J. Muscle. Res. Cell. Motil.* 19, 877–886.

Lindemann, C. B., Lesich, K. A. 2010. Flagellar and ciliary beating: The proven and the possible. *J. Cell. Sci.* 123, 519–528.

Llinas, P., Isabet, T., Song, L., Ropars, V., Zong, B., Benisty, H., Sirigu, S., Morris, C., Kikuti, C., Safer, D., Sweeney, H. L., Houdusse, A. 2015. How actin initiates the motor activity of myosin. *Dev. Cell.* 33, 401–412.

Lymn, R. W., Taylor, E. W. 1971. Mechanism of adenosine triphosphate hydrolysis by actomyosin. *Biochemistry* 10, 4617–4624.

Malnasi-Csizmadia, A., Dickens, J. L., Zeng, W., Bagshaw, C. R. 2005. Switch movements and the myosin crossbridge stroke. *J. Muscle Res. Cell Motil.* 26, 31–37.

Malnasi-Csizmadia, A., Kovacs, M. 2010. Emerging complex pathways of the actomyosin powerstroke. *Trends Biochem. Sci.* 35, 684–690.

Mehta, A. D., Rock, R. S., Rief, M., Spudich, J. A., Mooseker, M. S., Cheney, R. E. 1999. Myosin-V is a processive actin-based motor. *Nature* 400, 590–593.

Moore, J. R., Krementsova, E. B., Trybus, K. M., Warshaw, D. M. 2004. Does the myosin V neck region act as a lever? *J. Muscle. Res. Cell. Motil.* 25, 29–35.

Muretta, J. M., Petersen, K. J., Thomas, D. D. 2013. Direct real-time detection of the actin-activated power stroke within the myosin catalytic domain. *Proc. Natl. Acad. Sci. USA* 110, 7211–7216.

Muretta, J. M., Rohde, J. A., Johnsrud, D. O., Cornea, S., Thomas, D. D. 2015. Direct real-time detection of the structural and biochemical events in the myosin power stroke. *Proc. Natl. Acad. Sci. USA* 112, 14272–14277.

Nesmelov, Y. E., Agafonov, R. V., Negrashov, I. V., Blakely, S. E., Titus, M. A., Thomas, D. D. 2011. Structural kinetics of myosin by transient time-resolved FRET. *Proc. Natl. Acad. Sci. USA* 108, 1891–1896.

Nyitrai, M., Geeves, M. A. 2004. Adenosine diphosphate and strain sensitivity in myosin motors. *Philos. Trans. R. Soc. Lond. B. Biol. Sci.* 359, 1867–1877.

Onishi, H., Ohki, T., Mochizuki, N., Morales, M. F. 2002. Early stages of energy transduction by myosin: Roles of Arg in switch I, of Glu in switch II, and of the salt-bridge between them. *Proc. Natl. Acad. Sci. USA* 99, 15339–15344.

Ovchinnikov, V., Trout, B. L., Karplus, M. 2010. Mechanical coupling in myosin V: A simulation study. *J. Mol. Biol.* 395, 815–833.

Petit, C., Richardson, G. P. 2009. Linking genes underlying deafness to hair-bundle development and function. *Nat. Neurosci.* 12, 703–710.

Quintero, O. A., Moore, J. E., Unrath, W. C., Manor, U., Salles, F. T., Grati, M., Kachar, B., Yengo, C. M. 2010. Intermolecular autophosphorylation regulates myosin IIIa activity and localization in parallel actin bundles. *J. Biol. Chem.* 285, 35770–35782.

Rayment, I., Holden, H. M., Whittaker, M., Yohn, C. B., Lorenz, M., Holmes, K. C., Milligan, R. A. 1993a. Structure of the actin-myosin complex and its implications for muscle contraction. *Science* 261, 58–65.

Rayment, I., Rypniewski, W. R., Schmidt-Base, K., Smith, R., Tomchick, D. R., Benning, M. M., Winkelmann, D. A., Wesenberg, G., Holden, H. M. 1993b. Three-dimensional structure of myosin subfragment-1: A molecular motor. *Science* 261, 50–58.

Rayment, I., Smith, C., Yount, R. G. 1996. The active site of myosin. *Annu. Rev. Physiol.* 58, 671–702.

Roberts, A. J., Kon, T., Knight, P. J., Sutoh, K., Burgess, S. A. 2013. Functions and mechanics of dynein motor proteins. *Nat. Rev. Mol. Cell. Biol.* 14, 713–726.

Rosenfeld, S. S., Houdusse, A., Sweeney, H. L. 2005. Magnesium regulates ADP dissociation from myosin V. *J. Biol. Chem.* 280, 6072–6079.

Sakamoto, T., Wang, F., Schmitz, S., Xu, Y., Xu, Q., Molloy, J. E., Veigel, C., Sellers, J. R. 2003. Neck length and processivity of myosin V. *J. Biol. Chem.* 278, 29201–29207.

Sasaki, N., Shimada, T., Sutoh, K. 1998. Mutational analysis of the switch II loop of Dictyostelium myosin II. *J. Biol. Chem.* 273, 20334–20340.

Sellers, J. R., Veigel, C. 2006. Walking with myosin V. *Curr. Opin. Cell. Biol.* 18, 68–73.

Spudich, J. A., Aksel, T., Bartholomew, S. R., Nag, S., Kawana, M., Yu, E. C., Sarkar, S. S. et al. 2016. Effects of hypertrophic and dilated cardiomyopathy mutations on power output by human beta-cardiac myosin. *J. Exp. Biol.* 219, 161–167.

Sun, M., Rose, M. B., Ananthanarayanan, S. K., Jacobs, D. J., Yengo, C. M. 2008. Characterization of the pre-force-generation state in the actomyosin cross-bridge cycle. *Proc. Natl. Acad. Sci. USA* 105, 8631–8636.

Sweeney, H. L., Houdusse, A. 2010. Structural and functional insights into the myosin motor mechanism. *Annu. Rev. Biophys.* 39, 539–557.

Swenson, A. M., Trivedi, D. V., Rauscher, A. A., Wang, Y., Takagi, Y., Palmer, B. M., Malnasi-Csizmadia, A., Debold, E. P., Yengo, C. M. 2014. Magnesium modulates actin binding and ADP release in myosin motors. *J. Biol. Chem.* 289, 23977–23991.

Trivedi, D. V. 2014. Allosteric communication and force generation in myosin motors. Doctoral dissertation. Pennsylvania State University, University Park, PA.

Trivedi, D. V., Muretta, J. M., Swenson, A. M., Davis, J. P., Thomas, D. D., Yengo, C. M. 2015. Direct measurements of the coordination of lever arm swing and the catalytic cycle in myosin V. *Proc. Natl. Acad. Sci. USA.* 112, 14593–14598.

Vale, R. D., Milligan, R. A. 2000. The way things move: Looking under the hood of molecular motor proteins. *Science* 288, 88–95.

Volkmann, N., Liu, H., Hazelwood, L., Krementsova, E. B., Lowey, S., Trybus, K. M., Hanein, D. 2005. The structural basis of myosin V processive movement as revealed by electron cryomicroscopy. *Mol. Cell.* 19, 595–605.

Williams, D. S., Lopes, V. S. 2011. The many different cellular functions of MYO7A in the retina. *Biochem. Soc. Trans.* 39, 1207–1210.

Wulf, S. F., Ropars, V., Fujita-Becker, S., Oster, M., Hofhaus, G., Trabuco, L. G., Pylypenko, O., Sweeney, H. L., Houdusse, A. M., Schroder, R. R. 2016. Force-producing ADP state of myosin bound to actin. *Proc. Natl. Acad. Sci. USA.* 113, E1844–E1852.

Yengo, C. M., De La Cruz, E. M., Chrin, L. R., Gaffney, D. P. 2nd, Berger, C. L. 2002a. Actin-induced closure of the actin-binding cleft of smooth muscle myosin. *J. Biol. Chem.* 277, 24114–24119.

Yengo, C. M., De la Cruz, E. M., Safer, D., Ostap, E. M., Sweeney, H. L. 2002b. Kinetic characterization of the weak binding states of myosin V. *Biochemistry* 41, 8508–8517.

Yildiz, A., Forkey, J. N., McKinney, S. A., Ha, T., Goldman, Y. E., Selvin, P. R. 2003. Myosin V walks hand-over-hand: Single fluorophore imaging with 1.5-nm localization. *Science* 300, 2061–2065.

Zorca, C. E., Kim, L. K., Kim, Y. J., Krause, M. R., Zenklusen, D., Spilianakis, C. G., Flavell, R. A. 2015. Myosin VI regulates gene pairing and transcriptional pause release in T cells. *Proc. Natl. Acad. Sci. USA* 112, E1587–E1593.

chapter five

Distinguishing revolving motors from rotation motors

5.1 Distinct channel chirality of revolving motors and rotation motors

Chirality is one of the criteria to distinguish revolving motors from rotation motors. Recently, it has been reported that motor channels (the connector) of SPP1 (Lebedev et al., 2007), T7 (Agirrezabala et al., 2005), HK97, P22 (Olia et al., 2011), and phi29 (Guasch et al., 2002) all adopt an antichiral arrangement between the left-handed motor connector subunits and the right-handed DNA helices during packaging (Schwartz et al., 2013; Zhao et al., 2013). The helices lining the channels formed by the 12 subunits of all of these phages tilted at a 30° angle relative to the vertical axis of the connector channel, resulting in an antichiral configuration with the right-handed dsDNA (Figure 5.1). Such an antichiral structure greatly facilitates the controlled motion of one-way packaging when dsDNA revolves through the 12 subunits of the connector channel. The genome advances through revolving motion with no rotation, coiling, or torsion force by the contact between each of the 12 connector subunits, and the genome with a 30° transition for each of the 12 discrete steps (Schwartz et al., 2013). On the other hand, both the DNA strand and the helicase channel in rotation motors are right-handed with parallel threads.

5.2 Distinct channel size of revolving motors and rotation motors

Channel size is another criteria for distinction of revolving motors from rotation ones. It has been reported that rotation motors have a relatively narrow channel with a single strand of dsDNA inside the channel, to ensure full contact for DNA during its translocation through the center of the channel. As a result, the channel size is similar to or smaller than the diameter of dsDNA, which is 2 nm. In the case of revolving biomotors, however, both strands of dsDNA translocate the motor channel, and its diameter is visibly wider than that of rotation motors to ensure sufficient room for the substrate to revolve. dsDNA advances by its contact with the channel wall from the side instead of proceeding through the

(a)
Revolving motor

(b)
Rotation motor

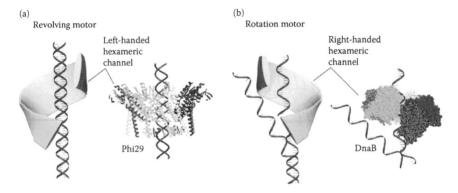

Left-handed
hexameric
channel

Right-handed
hexameric
channel

Phi29

DnaB

Figure 5.1 Chirality comparison of revolving and rotation biomotors. (a) In revolving motors, the right-handed DNA revolves within an anti-chiral left-handed channel. (b) In rotation motors, the right-handed DNA rotates through a right-handed channel via parallel threads (PDB ID: phi29-gp10, 1H5W; DnaB, 4ESV). (Adapted from De-Donatis, G. et al., 2014. *Cell Biosci.* 4, 30, with permission.)

center of the channel. Such expectations are confirmed by the inspection of the crystal structures of different motors, revealing that the diameters of the revolving motor channels are larger than 3 nm, while those of the rotation motors are smaller than 2 nm (De-Donatis et al., 2014), as evidenced in revolving motors of phi29 (Guo et al., 1987), SPP1, T4, T7, HK97,

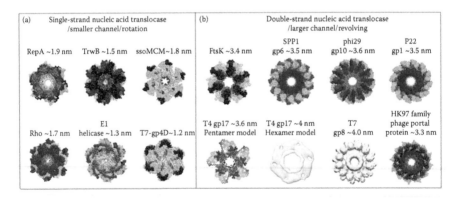

Figure 5.2 Comparison of the size of channels between biomotors. (a) Rotation motors use smaller channels (<2 nm), while (b) revolving motors use larger channels (>3 nm in diameter) for substrate translocation. (PDB codes: RepA, 1G8Y; TrwB: 1E9R.; ssoMCM, 2VL6; Rho, 3ICE.; E1, 2GXA; T7-gp4D, 1E0J; FtsK, 2IUU.; phi29-gp10, 1H5W; HK97 family-portal protein, 3KDR; SPP1-gp6, 2JES; P22-gp1, 3LJ5; T7-gp8 EM ID: EMD-1231.) (Adapted from De-Donatis, G. et al., 2014. *Cell Biosci.* 4, 30, with permission.)

and FtsK with a channel of 3–5 nm and rotation motors of Rho factor, E1 helicase, TrwB, ssoMCM, and RepA with a channel of 1–2 nm (Figure 5.2) (Valpuesta and Carrascosa, 1994; Agirrezabala et al., 2005; Massey et al., 2006; Lebedev et al., 2007; Sun et al., 2008).

The fact that in revolving motors, dsDNA advances by touching the channel wall instead of proceeding through the center of the channel (Guo et al., 2013) is consistent with recent Cryo-EM images, showing that the T7 dsDNA core tilts relatively to its channel axis (Figure 5.3). A counterclockwise motion of the dsDNA in the T7 motor was observed from the N-terminus of its connector (Guo et al., 2013), in agreement with the direction of dsDNA revolving in the phi29 dsDNA packaging motor (Jing et al., 2010; Zhang et al., 2012; Schwartz et al., 2013; Zhao et al., 2013).

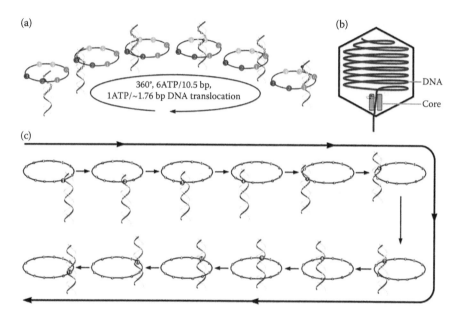

Figure 5.3 Model of sequence action of phi29 DNA packaging motor. Binding of ATP to the conformationally-disordered ATPase subunit stimulates an entropic and conformational change of the ATPase, thus fastening the ATPase at a less random configuration. This lower entropy conformation enables the ATPase subunit to bind dsDNA and prime ATP hydrolysis. ATP hydrolysis triggers the second entropic and conformational change, which renders the ATPase into a low affinity for dsDNA thus pushing the DNA to the next subunit that has already bound ATP. These sequential actions promote the movement and revolving of the dsDNA around the hexameric ATPase ring. (Adapted from Schwartz et al., 2013. *Virology* 443, 28–39; Guo, F. et al., 2013. *Proc. Natl. Acad. Sci. USA* 110, 6811–6816, with permission.)

References

Agirrezabala, X., Martin-Benito, J., Valle, M., Gonzalez, J. M., Valencia, A., Valpuesta, J. M., Carrascosa, J. L. 2005. Structure of the connector of bacteriophage T7 at 8A resolution: Structural homologies of a basic component of a DNA translocating machinery. *J. Mol. Biol.* 347, 895–902.

De-Donatis, G., Zhao, Z., Wang, S., Huang, P. L., Schwartz, C., Tsodikov, V. O., Zhang, H., Haque, F., Guo, P. 2014. Finding of widespread viral and bacterial revolution dsDNA translocation motors distinct from rotation motors by channel chirality and size. *Cell Biosci.* 4, 30.

Guasch, A., Pous, J., Ibarra, B., Gomis-Ruth, F. X., Valpuesta, J. M., Sousa, N., Carrascosa, J. L., Coll, M. 2002. Detailed architecture of a DNA translocating machine: The high-resolution structure of the bacteriophage phi29 connector particle. *J. Mol. Biol.* 315, 663–676.

Guo, F., Liu, Z., Vago, F., Ren, Y., Wu, W., Wright, E. T., Serwer, P., Jiang, W. 2013. Visualization of uncorrelated, tandem symmetry mismatches in the internal genome packaging apparatus of bacteriophage T7. *Proc. Natl. Acad. Sci. USA* 110, 6811–6816.

Guo, P., Peterson, C., Anderson, D. 1987. Prohead and DNA-gp3-dependent ATPase activity of the DNA packaging protein gp16 of bacteriophage phi29. *J. Mol. Biol.* 197, 229–236.

Jing, P., Haque, F., Shu, D., Montemagno, C., Guo, P. 2010. One-way traffic of a viral motor channel for double-stranded DNA translocation. *Nano Lett.* 10, 3620–3627.

Lebedev, A. A., Krause, M. H., Isidro, A. L., Vagin, A. A., Orlova, E. V., Turner, J., Dodson, E. J., Tavares, P., Antson, A. A. 2007. Structural framework for DNA translocation via the viral portal protein. *EMBO J.* 26, 1984–1994.

Massey, T. H., Mercogliano, C. P., Yates, J., Sherratt, D. J., Lowe, J. 2006. Double-stranded DNA translocation: Structure and mechanism of hexameric FtsK. *Mol. Cell* 23, 457–469.

Olia, A. S., Prevelige, P. E., Johnson, J. E., Cingolani, G. 2011. Three-dimensional structure of a viral genome-delivery portal vertex. *Nat. Struct. Mol. Biol.* 18, 597–603.

Schwartz, C., De Donatis, G. M., Zhang, H., Fang, H., Guo, P. 2013. Revolution rather than rotation of AAA+ hexameric phi29 nanomotor for viral dsDNA packaging without coiling. *Virology* 443, 28–39.

Sun, S., Kondabagil, K., Draper, B., Alam, T. I., Bowman, V. D., Zhang, Z., Hegde, S., Fokine, A., Rossmann, M. G., Rao, V. B. 2008. The structure of the phage T4 DNA packaging motor suggests a mechanism dependent on electrostatic forces. *Cell* 135, 1251–1262.

Valpuesta, J. M., Carrascosa, J. 1994. Structure of viral connectors and their funciton in bacteriophage assembly and DNA packaging. *Quart. Rev. Biophys.* 27, 107–155.

Zhang, H., Schwartz, C., De Donatis, G. M., Guo, P. 2012. "Push through one-way valve" mechanism of viral DNA packaging. *Adv. Virus Res.* 83, 415–465.

Zhao, Z., Khisamutdinov, E., Schwartz, C., Guo, P. 2013. Mechanism of one-way traffic of hexameric phi29 DNA packaging motor with four electropositive relaying layers facilitating anti-parallel revolution. *ACS Nano* 7, 4082–4092.

chapter six

General mechanism of biomotors

6.1   Force generation and energy conversion

In all biomotors, the nucleotide binding and hydrolysis cycles are coupled to conformational entropy rearrangements of substrate-binding subunits (Kolomeisky and Fisher, 2007). Till date, there are three primary chemo-mechanical coupling models for biomotors: sequential (individual ATP-binding/hydrolysis events proceed sequentially), concerted (all active sites hydrolyze ATP simultaneously), and stochastic (any ATPase site can hydrolyze nucleotide randomly) models. In general, binding of ATP to the disordered ATPase subunit stimulates a conformational change with entropy alteration (De-Donatis et al., 2014) of the ATPase, thus fastening the ATPase at a less random configuration. This new conformation entropy enables the ATPase subunit to bind dsDNA and prime ATP hydrolysis. ATP hydrolysis triggers the second entropic and conformational change, which renders the ATPase into a low affinity for dsDNA and, thus, pushes the DNA to the next subunit that has already bound to ATP with a high affinity for dsDNA. These continuous actions will promote the movement of the dsDNA around the internal ring.

6.2   Motor subunit communication

The signal transfer mode among motor subunits in different ring-trans-locases is not identical. Sequential action of the phi29 dsDNA packaging motor was originally reported by Chen and Guo (1997) and subsequently confirmed by Moffitt and coworkers (2009). Hill constant determination and binomial distribution of inhibition assay have led to the conclusion that ATPase subunits work sequentially and cooperatively (Schwartz et al., 2013; De-Donatis et al., 2014). This action enables the motor to work continuously without interruption, despite some observable pauses. Extensive studies towards the study of motor subunit communications draw researchers' attention to the arginine finger motif in the biological motors.

Functions of arginine finger motifs slightly vary among different ATPases. The overall structural features of the AAA+ core domains are conserved in all ATPases of the superfamily with a highly conserved arginine residue close to the sensor 2. The helicase superfamily III proteins possess an aberrantly formed α-helical domain, thus they lack the sensor 2 arginine. The polar residues in sensor 1 were reported to mediate conformational changes, though not all AAA+ proteins possess a polar residue in the sensor 1 position. In the case of p97 D2, the movement of the sensor 1 residue upon nucleotide engagement in the binding pocket induces displacements at the distal end of the ASCE domain, where the arginine fingers are located. The conserved arginine in the α-helical subdomain was termed sensor 2 residue, which contacts the phosphate groups of ATP and mediate a conformational change that sequesters the catalytic site from water. Mutations of the sensor 2 residues lead to a loss or decrease of ATP binding and/or ATP hydrolysis (Hanson and Whiteheart, 2005; Elles and Uhlenbeck, 2008; Chen et al., 2010). They also seem to be important for the stability of hexameric complex, since arginine finger mutations in phi29 gp16 ATPase, HslU, p97 VCP, ClpB D1, ClpC D1, and Hsp104 D1 impair oligomer formation even in the presence of ATP (Hanson and Whiteheart, 2005). Hence, the residues not only participate in hydrolysis but also account for subunit interface contacts and/or the structural integrity of the binding pocket (Hanson and Whiteheart, 2005; Elles and Uhlenbeck, 2008; Chen et al., 2010); in other words, motor subunit communication.

Taking the most recent report about phi29 gp16 ATPase as an example, the arginine finger in the ATPase was aligned and mutated (Zhao et al., 2016, 2017). The arginine finger-free ATPase has shown abolished activities in ATP binding/hydrolysis, nucleotide binding, oligomerization, and *in vitro* virion assembly. Ultracentrifugation assays and electrophoresis mobility shift assays (EMSA) have demonstrated that arginine finger in gp16 ATPase extends from one subunit to the adjacent one and bridges dimer formation (Figure 6.1) (Zhao et al., 2016, 2017). Arginine mutants alone could not form dimers, while interactions were observed when they were mixed with either the wild-type or other mutants that contained an intact arginine finger, which can provide an arginine residue for dimer formation (Figure 6.1). Interestingly, the isolated dimer alone does not display any assembly activity (Figure 6.1), agreeing with the previous observation that fresh monomers have to be added into the packaging intermediate in order to reinitiate the packaging process (Shu and Guo, 2003; Zhao et al., 2016, 2017). Most hexameric AAA+ structures show a typical domain arrangement, in which the nucleotide binding pocket lies in the interface between two protomers. Such a structural arrangement supports the conclusion that in the active complex ring, the arginine finger of one subunit comes into close proximity to the nucleotide bound in the neighboring subunit (Figure 6.2). A hexameric ring of phi29 gp16 ATPase has also been modeled and aligned with that of the hexameric FtsK DNA translocase of *Escherichia coli*. The position of the arginine finger of one subunit of gp16 is shown to stretch to the active site of a neighboring subunit (Figure 6.3), agreeing with the other ATPases that arginine finger was part of the ATP-binding pocket for cooperative behavior.

Arginine finger involved in ATPase intersubunit interaction

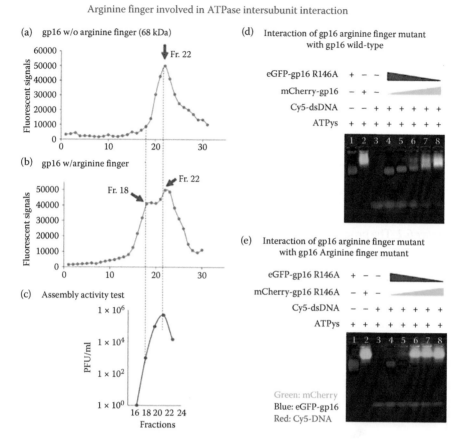

Figure 6.1 Inter-subunit communication of gp16 ATPase mediated by arginine finger motif. (a–c) Ultracentrifugation assay showing the presence of both dimers (showing no biological activity) and monomers in the gp16 ATPase rings. (d–e) EMSA between gp16 arginine finger mutants with gp16 wild-type (d) or with arginine finger mutants (e), showing the bridging effect of arginine finger between two adjacent subunits. (Adapted from Zhao, Z. et al., 2016. *Mol. Cell. Biol.* 36, 2514–2523; Zhao, Z. et al., 2017. *Small* 13 with permission.)

With the above evidence, the cooperativity of subunit communication regulated by arginine finger has been illustrated. Oligomeric helicases contain one ASCE domain per monomer, with the ATP site at the interface between adjacent subunits, and rely on the interaction with neighboring subunits to provide the full nucleotide-binding pocket. In contrast, ATPases from SF1 and SF2 (Superfamily 1 and Superfamily 2) typically contain tandem ASCE folds and bind the nucleotide at the interface between the two domains, with the N-terminal domain providing

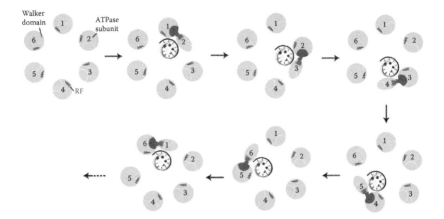

Figure 6.2 The proposed mechanism of ATPase coordination with a series of conformational changes during DNA binding and ATP hydrolysis that are regulated by the arginine finger (RF, red). (Adapted from Zhao, Z. et al., 2016. *Mol. Cell. Biol.* 36, 2514–2523, with permission.)

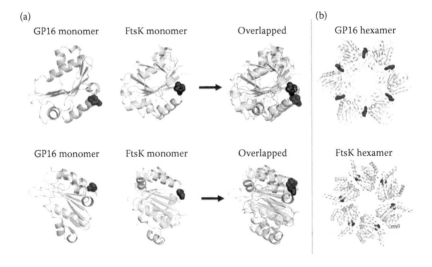

Figure 6.3 Prediction and comparison of gp16 structure. (a) Structural comparison between the crystal structure of FtsK monomer (PDB accession number 2IUU; cyan) and the gp16 ATPase model (pink). The arginine finger is highlighted as a sphere. (b) Comparison of the predicted gp16 hexamer and FtsK hexamer. The ATP domains are highlighted as spheres: residue 27 (green, the conserved Walker ATP domain) and residue 146 (red, the arginine finger). (Adapted from Zhao, Z. et al., 2016. *Mol. Cell. Biol.* 36, 2514–2523, with permission.)

the Walker A and Walker B motif, and the C-terminal providing other elements. The homologous type II transmembrane proteins LAP1 and LULL1 adopt nucleotide-free ATPase folds, and donate arginine fingers to complete the active sites of Torsin ATPases. The ring structure provides a central channel where the nucleic acid substrate is supposed to thread.

A closer look at the GTP binding site of the crystallized Ras/p120GAP/GDP AlF3 complex reveals that the closest distance between the fluoride groups of AlF3 and the amino group of Arg789 is around 2.6 Å, allowing for direct interactions between the arginine finger and the nucleotide in the transition state. The ATP-binding pocket of the SV40 LTag helicase (helicase superfamily III) is formed by two positively charged residues from the neighboring subunit, namely Arg540 and Lys418. Interestingly, Lys418, rather than the conserved arginine finger residue Arg540, takes the location of that in the Ras/p120GAP pair (Wendler et al., 2012).

As demonstrated above, arginine finger serves as a bridge between two independent subunits, thus forming a transient dimeric subunit. In wild-type gp16 ATPase, it was observed that both dimer and monomer forms were present in solution, as revealed by glycerol gradient centrifugation experiments (Zhao et al., 2016, 2017). The communication between each two adjacent subunits mediated by the arginine finger results in an asymmetrical hexameric organization, which is supported by the asymmetrical structure in many other hexameric ATPase systems as shown by structural computation, X-ray diffraction and Cryo-EM imaging (Lyubimov et al., 2012; Arai et al., 2013; Stinson et al., 2015; Ye et al., 2015;

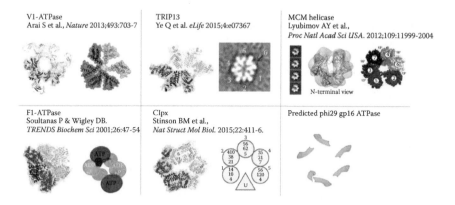

Figure 6.4 Asymmetrical structures of various ATPase hexamers. Structures of V1-ATPase, TRIP13, MCM helicase, ClpX are shown as representatives. PDB ID: V1-ATPase, 3VR5; TRIP13, 4XGU; F1-ATPase, 1BMF; ClpX, 4I81. (Adapted from Zhao, Z. et al., 2016. *Mol. Cell. Biol.* 36, 2514–2523; Stinson, B. M. et al., 2015. *Nat. Struct. Mol. Biol.* 22, 411–416; Lyubimov, A. Y. et al., 2012. *Proc. Natl. Acad. Sci. USA* 109, 11999–12004; Ye, Q. et al., 2015. *Elife* 4; Arai, S. et al., 2013. *Nature* 493, 703–707, with permission.)

Zhao et al., 2016) (Figure 6.4). This phenomenon could provide some clues as to why the asymmetrical hexameric ATPase of gp16 of phi29 and gp17 of T4 was previously interpreted as having a pentameric configuration by cryo-EM. Since the two adjacent subunits of the ATPase could interact with each other and form a closer dimer configuration, this dimer will appear as a monomeric subunit different from the other subunits and the hexameric ring as asymmetrical.

References

Arai, S., Saijo, S., Suzuki, K., Mizutani, K., Kakinuma, Y., Ishizuka-Katsura, Y. et al. 2013. Rotation mechanism of *Enterococcus hirae* V1-ATPase based on asymmetric crystal structures. *Nature* 493, 703–707.

Chen, B., Sysoeva, T. A., Chowdhury, S., Guo, L., De, C. S., Hanson, J. A., Yang, H., Nixon, B. T. 2010. Engagement of arginine finger to ATP triggers large conformational changes in NtrC1 AAA+ ATPase for remodeling bacterial RNA polymerase. *Structure* 18, 1420–1430.

Chen, C., Guo, P. 1997. Sequential action of six virus-encoded DNA-packaging RNAs during phage phi29 genomic DNA translocation. *J. Virol.* 71, 3864–3871.

De-Donatis, G., Zhao, Z., Wang, S., Huang, P. L., Schwartz, C., Tsodikov, V. O., Zhang, H., Haque, F., Guo, P. 2014. Finding of widespread viral and bacterial revolution dsDNA translocation motors distinct from rotation motors by channel chirality and size. *Cell Biosci.* 4, 30.

Elles, L. M., Uhlenbeck, O. C. 2008. Mutation of the arginine finger in the active site of *Escherichia coli* DbpA abolishes ATPase and helicase activity and confers a dominant slow growth phenotype. *Nucleic Acids Res.* 36, 41–50.

Hanson, P. I., Whiteheart, S. W. 2005. AAA+ proteins: Have engine, will work. *Nat. Rev. Mol. Cell. Biol.* 6, 519–529.

Kolomeisky, A. B., Fisher, M. E. 2007. Molecular motors: A theorist's perspective. *Annu.Rev.Phys.Chem.* 58, 675–695.

Lyubimov, A. Y., Costa, A., Bleichert, F., Botchan, M. R., Berger, J. M. 2012. ATP-dependent conformational dynamics underlie the functional asymmetry of the replicative helicase from a minimalist eukaryote. *Proc. Natl. Acad. Sci. USA* 109, 11999–12004.

Moffitt, J. R., Chemla, Y. R., Aathavan, K., Grimes, S., Jardine, P. J., Anderson, D. L., Bustamante, C. 2009. Intersubunit coordination in a homomeric ring ATPase. *Nature* 457, 446–450.

Schwartz, C., De Donatis, G. M., Zhang, H., Fang, H., Guo, P. 2013. Revolution rather than rotation of AAA+ hexameric phi29 nanomotor for viral dsDNA packaging without coiling. *Virology* 443, 28–39.

Shu, D., Guo, P. 2003. Only one pRNA hexamer but multiple copies of the DNA-packaging protein gp16 are needed for the motor to package bacterial virus phi29 genomic DNA. *Virology* 309(1), 108–113.

Stinson, B. M., Baytshtok, V., Schmitz, K. R., Baker, T. A., Sauer, R. T. 2015. Subunit asymmetry and roles of conformational switching in the hexameric AAA+ ring of ClpX. *Nat. Struct. Mol. Biol.* 22, 411–416.

Wendler, P., Ciniawsky, S., Kock, M., Kube, S. 2012. Structure and function of the AAA+ nucleotide binding pocket. *Biochim. Biophys. Acta* 1823, 2–14.

Ye, Q., Rosenberg, S. C., Moeller, A., Speir, J. A., Su, T. Y., Corbett, K. D. 2015. TRIP13 is a protein-remodeling AAA+ ATPase that catalyzes MAD2 conformation switching. *Elife* 4.

Zhao, Z., De-Donatis, G. M., Schwartz, C., Fang, H., Li, J., Guo, P. 2016. Arginine finger regulates sequential action of asymmetrical hexameric ATPase in dsDNA translocation motor. *Mol. Cell. Biol.* 36, 2514–2523.

Zhao, Z., Zhang, H., Shu, D., Montemagno, C., Ding, B., Li, J., Guo, P. 2017. Construction of asymmetrical hexameric biomimetic motors with continuous single-directional motion by sequential coordination. *Small* 13.

chapter seven

Mechanism of revolving motors

7.1   Revolving motion in biological motors

ATPase is an enzyme that facilitates the hydrolysis of ATP, and gp16 in the phi29 dsDNA packaging motor converts the energy that results from this process into physical motion. The valve of the connector channel between the motor and procapsid is structured in such a way that it allows the dsDNA to enter the chamber but not exit. A virus-encoded packaging RNA, or pRNA, is required for the translocation of dsDNA into the phi29 procapsid. It coordinates the action of the separate components of gp16 and the central channel, making for an integrated motor. When ATP interacts with gp16, a conformational change in gp16 increases its binding to dsDNA. Finally, the hydrolysis of ATP functions to force dsDNA through the channel.

The FtsK motor domain from *Pseudomonas aeruginosa* is a sixfold symmetric ring, with ADP bound in every subunit (Massey et al., 2006) as revealed by crystallization studies. The same study also presented a monomeric crystal form of the motor domain with ATP bound. When the β domains of these two structures were superimposed, their α domains were shifted relative to each other in a hinge-like opening, which was able to move the point in the α domain juxtaposed to the DNA substrate by 5.5 Å (equivalent to 1.6 bp). The translocation of 1.6 bp of dsDNA per subunit of FtsK (Massey et al., 2006; Crozat and Grainge, 2010) might be correlated to ATP hydrolysis. It strongly agrees with the step size of the phi29 DNA packaging motor, in which one ATPase subunit uses one ATP each time to package 1.75 nucleotides (Guo et al., 1987, 2013; Schwartz et al., 2013a,b; Zhao et al., 2013), and also that in bacteriophage T3 with 1.8 bp per ATP (Morita et al., 1993). Based on that, it has been proposed that FtsK possesses a rotary inchworm mechanism (Massey et al., 2006; Crozat and Grainge, 2010), with ATPase subunits acting sequentially. In the model proposed by Massey et al., each hydrolyzed ATP moves the DNA by 2 bp ($\sim$6.8 Å, larger than the 5.5 Å difference between the two crystal forms as discussed above) in an inchworm-like movement with the central DNA having contact with both end domains. The relative strength of interaction between each domain and the DNA is dependent upon the ATP binding or hydrolysis state; upon ATP hydrolysis, one DNA contact is lost while the other contact is strong but shifted by $\sim$6.8 Å, resulting in net movement of 2 bp of DNA along the central channel (Guo et al., 2016). An integer number of bases were chosen as the most likely model as it allows each monomer in the ring to contact the dsDNA in the same manner at every subunit, whether this is with the repeating sugar–phosphate backbone or the bases themselves. Movement by a noninteger number of bases means that the protein–DNA contact necessary for movement of the substrate would be different in any two adjacent subunits (Guo et al., 2016). The movement of dsDNA by one subunit also functions to bring the next monomer in the ring into contact with the DNA, handing the DNA substrate on to the adjacent monomer with minimal rotation. The next subunit also translocates 2 bp of DNA, and so on around the ring. The result is that 12 bp of DNA is translocated per catalytic cycle of the hexamer where all six monomers hydrolyze ATP once. This figure is close to the 10.5 bp per helical turn in B-form DNA, with the extra 1.5 bp appearing as a slight rotation. If the protein is anchored (both FtsK and SpoIIIE are membrane bound at their N-termini), it results in the generation of positive DNA supercoiling ahead of the motor and negative behind. Indeed, induction of supercoiling has been seen in bulk biochemical assays (Aussel et al., 2002) and in single-molecule experiments (Saleh et al., 2005). The observed supercoiling induction for FtsK translocation of one positive supercoil ahead of the motor per 150 bp translocated is in broad agreement with the rotary inchworm model (Guo et al., 2016).

The rotary inchworm model adopted for the FtsK motor also proposed an obligatory hand off event between adjacent monomers within a single ring, such that the presence of a single catalytically inactive subunit would effectively inactivate the entire hexamer (Guo et al., 2016). This was backed up by biochemical data: Mutants that were unable to bind ATP were mixed in different ratios with wild-type subunits, and the relative ATPase activity was compared to wild type only. With an increasing amount of mutant subunits, ATPase activity decreased rapidly, confirming the predicted pattern (Guo et al., 2016).

Interestingly, a fusion protein has been produced in which three linked motor domains, effectively a covalent trimer of FtsK, were contained in a single polypeptide. This construct was found to be a very active DNA translocase motor (Crozat et al., 2010). Within it, the Walker A and Walker B motifs, for nucleotide binding and hydrolysis, respectively, could be mutated in specific subunits, leading to the surprising finding that a single active-site mutant, or two nonadjacent mutants per hexamer, did not cause a great decrease in ATPase activity or the speed of translocation along dsDNA. However, the presence of these mutations did reduce the ability of the hexamer to produce force as judged by the ability of the protein to displace either protein or DNA triplexes (Crozat et al., 2010). It is important to consider that when the ATPase subunits were fused into a concatemer, the ATPase could refold or form a higher order multimer, rendering the interpretation of the results challenging. In addition, the linking of subunits as a concatemer may relax the otherwise strict inter-subunit interactions that coordinate the binding and hydrolysis of ATP in the protein (such as the role of the arginine finger that stretches between one monomer and the active site of its neighbor).

In order to explain these results, a new model was proposed in which more than one subunit within a hexameric ring would contact DNA concurrently, based upon the escort or "spiral staircase" model of Rho and E1 helicases (Crozat et al., 2010). In these hexameric helicases, multiple subunits can contact the DNA/RNA substrate at the same time, with the single contact point for each monomer being at a different level around the ring, rather like the stairs in a spiral staircase. ATP hydrolysis forcefully moves one of the contact points downward through the ring, and the other contacts move along passively. When the last contact point at the bottom of the ring is reached, the protein arm becomes free and can then move back up the top to re-engage with the polynucleotide substrate and reinitiate the cycle of movement down the staircase (Guo et al., 2016). With a flexible and compressible single-stranded DNA/RNA substrate, the movement of the protein–DNA contacts is small enough that the protein can remain with the DNA/RNA through a full catalytic cycle of every subunit in the ring. However, dsDNA is a much stiffer and noncompressible substrate than ssDNA or RNA, so a DNA translocase would have immense trouble utilizing an identical mechanism. Within a DNA translocase channel, each single protein contact would have to move almost 30 Å to maintain contact with the DNA during a full catalytic cycle around the hexameric ring. This amount of movement seems unlikely (Guo et al., 2016). However, in the channel of the DNA packaging motor of all dsDNA bacteriophages, a 30° left-handed channel structure was found. Such a left-handed configuration might assist to compensate for such constraints during shifting of the dsDNA and maintain the balanced contact during the advance of the dsDNA. Furthermore, it would be energetically unfeasible to compress the double helix greatly yet still produce power from ATP hydrolysis. Therefore, an intermediate or "limited-escort" model was proposed for FtsK, whereby three adjacent monomers in a ring contact dsDNA simultaneously. This could allow for one inactive subunit to be skipped if the monomers on either side of it are able to translocate. It would also imply that two adjacent mutant subunits would produce an inactive hexamer, which is consistent with the data (Guo et al., 2016).

Recent results for SpoIIIE also support an escort model (Liu et al., 2015); using a single-molecule setup with an optical trap, it was found that SpoIIIE most likely has a step-size on DNA of 2 bp, and that (at least) two adjacent subunits must be able to contact DNA simultaneously (Liu et al., 2015). It is noteworthy that this model is based on hexameric SpoIIIE made of six monomers, not linked multimers as with the FtsK experiments above, removing the concern of how the linked multimers may form an active translocase. Together these studies suggest a highly coordinated cycle of ATP binding and hydrolysis around the ring, with monomers becoming active and hydrolyzing ATP sequentially. Yet the hexameric rings retain the ability to bypass individual inactive monomers by virtue of having multiple concurrent DNA contacts (two or three monomers contacting DNA at the same time).

Both the inchworm and partial escort models are largely consistent with a revolving motor mechanism. They propose that dsDNA touches the internal surface of the hexameric ring, and that the contact point between protein and DNA revolves around the inner surface of the protein multimer with minimal rotation (Guo et al., 2016). They also suggest that a defined number of bases would be moved per hydrolyzed ATP; if each subunit contacts the DNA substrate in an identical fashion around the ring, then the DNA must be moved by a defined length at each step to maintain constant interaction. A slight twisting of the DNA at each step is then necessary to maintain the contact with the dsDNA around the ring, preserving the register between DNA and protein. While the angle between adjacent subunit active sites in a hexamer is 360°/6 or 60°, the angle between adjacent phosphates around the dsDNA axis is 360°/10.5 (about 34°) (Guo et al., 2016). If precisely 2 bp are translocated per subunit, the DNA must twist an extra 8° to maintain the identity of each protein–DNA contact. DNA within cells is negatively supercoiled, with a supercoiling density in *E. coli* of around −0.05. Thus, the amount of twisting required for each FtsK power step is reduced to around 5° per 2 bp translocated, which corresponds to one supercoil induced for every 144 bp translocated (Guo et al., 2016). This theoretical value is almost identical to the value of one supercoil per 150 bp observed using single molecule experiments. If this is the case, small overwinding of the DNA ahead of a translocating FtsK will produce positive supercoiling ahead of the protein; however, if the left-handed configuration as observed in the DNA packaging motor of dsDNA bacteriophages exists, then the supercoiling will not have occurred. On the other hand, in the cell, this might be removed by either the action of DNA gyrase or the occasional slipping of the motor to release the torsional tension in the DNA (Guo et al., 2016).

Furthermore, SpoIIIE is hypothesized to gain directionality through recognition of skewed 8 bp sequences in an analogous manner to FtsK. The SpoIIIE recognition sequence (SRS) of *Bacillus subtilis* is also a skewed 8 bp sequence (GAGAAGGG) and is bound by the C-terminal domain. However, recent models propose a subtly different mechanism for ensuring that SpoIIIE translocates DNA in the desired direction: rather than SRS acting as a preferential loading site onto the DNA, as proposed for FtsK (Grainge et al., 2011). SpoIIIE may randomly associate with DNA and scan for SRS sequences without the motor ATPase being active by a sliding/hopping mechanism. It has also been suggested that the γ domain of SpoIIIE acts to inhibit motor activity (Besprozvannaya et al., 2013) and that subsequent encounter of the passively sliding/hopping motor with an SRS sequence in the correct orientation relieves this repression. An alternative hypothesis is that the encounter of SRS leads to a conformational change in SpoIIIE, converting the inactive "open hexamer" which is capable of rapidly dissociating from DNA to a stable closed form in which ATPase activity is now greatly activated (Cattoni et al., 2013, 2014). Both FtsK and SpoIIIE can bind to random DNA, and ATPase activity is not dependent upon the presence of KOPS/SRS. However, the interaction between the motor protein and its cognate DNA sequence leads to motor activation by loading a hexamer, stabilizing a closed form of the hexamer on the DNA, relieving motor inhibition, or perhaps by a combination of these mechanisms. Whether FtsK and SpoIIIE in fact respond to their respective directionality sequences differently or the two related proteins behave in a similar fashion, remains to be seen.

7.2   One-way traffic of revolving biomotors

As reported, revolving biomotors drive their substrate translocation in a one-way traffic manner under the coordination of the motor components. Several factors contribute to the directional translocation: The loop and the four lysine rings aligned within the connector inner wall; ATP coupled conformational entropy alternations; left-handed configuration of the connector subunit; and the recognition of 5′–3′ strand for the single-directional movement.

The one-way flow loop within the connector channel of the phi29 dsDNA packaging motor is one of the most important factors determining the advancement of genome toward the capsid without reversing (Figure 7.1) (Fang et al., 2012; Geng et al., 2013; Zhao et al., 2013). Both biochemical and biophysical experiments with loop-deleted connectors have showed the ability of two-way traffic of the DNA substrate (Isidro et al., 2004; Serwer, 2010; Geng et al., 2011; Grimes et al., 2011; Fang et al., 2012), proving the role of these inner loops as a ratchet or clamp during DNA translocation, in line with the "push-through one-way valve" model (Jing et al., 2010; Fang et al., 2012; Zhang et al., 2012; Zhao et al., 2013). Structural studies of the SPP1 bacteriophage connector channel have shown the close proximity of its inner loops to the dsDNA substrate via nonionic interactions (Orlova et al., 2003) and the high similarity between SPP1 and phi29 protein channels (Chai et al., 1994), indicating the conserved role of the loops during genome packaging. Besides, the conserved 30° left-handed twist of the channel, which is antichiral to the right-handed dsDNA substrate, will compensate for the angle and distance difference by each contact between dsDNA and the channel, supporting the one-way revolving motion without rotation.

During packaging, the dsDNA genome is processed by contact of the connector with one strand of DNA in the 5′–3′ direction, as modification of

(a)

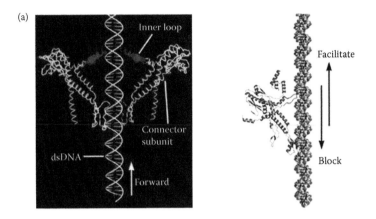

(b) Connector with inner loop for dsDNA – one way traffic

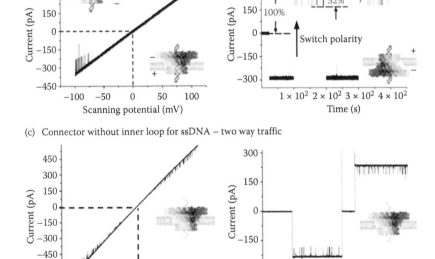

(c) Connector without inner loop for ssDNA – two way traffic

Figure 7.1 Role of the inner channel loop of phage portal proteins determines DNA one-way traffic. (a) Illustration of flexible loops within the phi29 (left) and SPP1 (right) connector channels. (b) Demonstration of one-way traffic of DNA through wild-type connectors using a ramping potential or a switching polarity (right). (c) Two-way traffic of DNA with a loop-deleted connector. (Adapted from Zhao, Z. et al., 2013. *ACS Nano* 7, 4082–4092, with permission.)

only that strand abolished dsDNA packaging (Oram et al., 2008; Aathavan et al., 2009). Extensions up to 12 bases at the end is tolerable, while extensions to 20 or more bases significantly blocked the DNA packaging of the T4 motor, agreeing with the previous conclusion since one complete circle of revolving is equal to one helical turn of 10.5 bp of dsDNA. In addition, force generation from ATPase also contributes to the one-way movement. As described in the early section, ATP binding and hydrolysis induce two consecutive entropic and conformational alternations of ATPase with either a high or low affinity for DNA, with the arginine finger motif as a mediator for signal transfer, which results in a sequential action facilitating the translocation of the DNA substrate in one direction.

Four lysine residues are reported to be aligned within each subunit of the phi29 connector interior surface (Guasch et al., 2002) (Figure 7.2). Similar patterns have also been observed in the phages of SPP1, P22, and phi29 (Simpson et al., 2000; Guasch et al., 2002). These electropositive

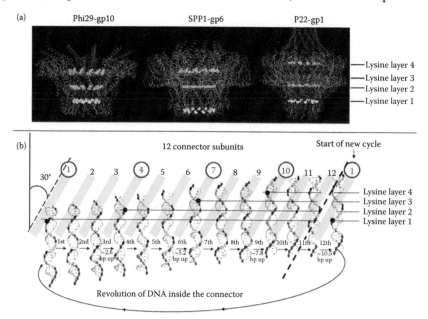

Figure 7.2 Four lysine rings facilitate one-way substrate traffic and result in four steps of pauses during each translocation cycle. (a) The presence of four positively charged lysine rings in different motors. (b) Diagram showing DNA revolving inside phi29 connector channel with four steps of pauses due to the interaction of four positively charged lysine rings with the negatively charged dsDNA phosphate backbone. (PDB codes: phi29-gp10, 1H5W; SPP1-gp6, 2JES; P22-gp1, 3LJ5). (Adapted from Zhao, Z. et al., 2013. *ACS Nano* 7, 4082–4092 and Guo, P. et al. 2014. *Biotechnology Advances* 32, 853–872, with permission.)

lysines form four relaying rings and interact with the electronegative DNA phosphate backbone, generating an auxiliary force involved in genome packaging (Fang et al., 2012; Schwartz et al., 2013a,b; Zhao et al., 2013). This resulting force leads to uneven speed alternations during the DNA translocation, mostly with four pauses (Chistol et al., 2012; Schwartz et al., 2013b; Zhao et al., 2013), as previously reported in both phi29 (Moffitt et al., 2009, Chistol et al., 2012) and T4 (Kottadiel et al., 2012).

The effects of the lysine layers on genome translocation can be interpreted using phi29 biomotor as an example (Zhao et al., 2013). Vertically, these four lysine layers of phi29 fall within 3.7 nm (Guasch et al., 2002) inside the 7-nm connector channel, that is an average of $\sim$0.9 nm apart between each two rings. During the revolution of DNA inside the motor channel, the distance variation due to the mismatch between genome base (10.5 bp for 360°) and the channel subunits (12 subunits) can be compensated by interactions between dsDNA phosphate backbone and the positively charged lysine from the next subunit, leading to a slight pause in DNA advancement. Continuation of the interactions lead to the four pauses during packaging as mentioned above.

References

Aathavan, K., Politzer, A. T., Kaplan, A., Moffitt, J. R., Chemla, Y. R., Grimes, S., Jardine, P. J., Anderson, D. L., Bustamante, C. 2009. Substrate interactions and promiscuity in a viral DNA packaging motor. *Nature* 461, 669–673.

Aussel, L. et al. 2002. FtsK Is a DNA motor protein that activates chromosome dimer resolution by switching the catalytic state of the XerC and XerD recombinases. *Cell* 108(2), 195–205.

Besprozvannaya, M. et al. 2013. SpoIIIE protein achieves directional DNA translocation through allosteric regulation of ATPase activity by an accessory domain. *J Biol Chem* 288(40), 28962–28974.

Cattoni, D. I. et al. 2013. SpoIIIE mechanism of directional translocation involves target search coupled to sequence-dependent motor stimulation. *EMBO Rep* 14(5), 473–479.

Cattoni, D. I. et al. 2014. Structure and DNA-binding properties of the Bacillus subtilis SpoIIIE DNA translocase revealed by single-molecule and electron microscopies. *Nucleic Acids Res* 42(4), 2624–2636.

Chai, S., Kruft, V., Alonso, J. C. 1994. Analysis of the *Bacillus subtilis* bacteriophages SPP1 and SF6 gene 1 product: A protein involved in the initiation of headful packaging. *Virology* 202, 930–939.

Chistol, G., Liu, S., Hetherington, C. L., Moffitt, J. R., Grimes, S., Jardine, P. J., Bustamante, C. 2012. High degree of coordination and division of labor among subunits in a homomeric ring ATPase. *Cell* 151, 1017–1028.

Crozat, E., Grainge, I. 2010. FtsK DNA translocase: The fast motor that knows where it's going. *Chembiochem* 11, 2232–2243.

Fang, H., Jing, P., Haque, F., Guo, P. 2012. Role of channel lysines and "push through a one-way valve" mechanism of viral DNA packaging motor. *Biophys. J.* 102, 127–135.

Geng, J., Fang, H., Haque, F., Zhang, L., Guo, P. 2011. Three reversible and controllable discrete steps of channel gating of a viral DNA packaging motor. *Biomaterials* 32, 8234–8242.

Geng, J., Wang, S., Fang, H., Guo, P. 2013. Channel size conversion of phi29 DNA-packaging nanomotor for discrimination of single- and double-stranded nucleic acids. *ACS Nano* 7, 3315–3323.

Grainge, I. et al. 2011. Activation of XerCD-dif recombination by the FtsK DNA translocase. *Nucleic Acids Res* 39(12), 5140–5148.

Grimes, S., Ma, S., Gao, J., Atz, R., Jardine, P. J. 2011. Role of phi29 connector channel loops in late-stage DNA packaging. *J. Mol. Biol.* 410, 50–59.

Guasch, A., Pous, J., Ibarra, B., Gomis-Ruth, F. X., Valpuesta, J. M., Sousa, N., Carrascosa, J. L., Coll, M. 2002. Detailed architecture of a DNA translocating machine: The high-resolution structure of the bacteriophage phi29 connector particle. *J. Mol. Biol.* 315, 663–676.

Guo, P. et al. 2014. Common mechanisms of DNA translocation motors in bacteria and viruses using one-way revolution mechanism without rotation. *Biotechnology Advances* 32, 853–872.

Guo, P., Noji, H., Yengo, C. M., Zhao, Z., Grainge, I. 2016. Biological nanomotors with revolution, linear, or rotation motion mechanism. *Microbiol. Mol. Biol. Rev.* 80, 161–186.

Guo, P., Peterson, C., Anderson, D. 1987. Prohead and DNA-gp3-dependent ATPase activity of the DNA packaging protein gp16 of bacteriophage phi29. *J. Mol. Biol.* 197, 229–236.

Guo, P., Schwartz, C., Haak, J., Zhao, Z. 2013. Discovery of a new motion mechanism of biomotors similar to the earth revolving around the sun without rotation. *Virology* 446, 133–143.

Isidro, A., Henriques, A. O., Tavares, P. 2004. The portal protein plays essential roles at different steps of the SPP1 DNA packaging process. *Virology.* 322, 253–263.

Jing, P., Haque, F., Shu, D., Montemagno, C., Guo, P. 2010. One-way traffic of a viral motor channel for double-stranded DNA translocation. *Nano Lett.* 10, 3620–3627.

Kottadiel, V. I., Rao, V. B., Chemla, Y. R. 2012. The dynamic pause-unpackaging state, an off-translocation recovery state of a DNA packaging motor from bacteriophage T4. *Proc. Natl. Acad. Sci. USA* 109, 20000–20005.

Liu, N. et al. 2015. Two-subunit DNA escort mechanism and inactive subunit bypass in an ultra-fast ring ATPase. *Elife* 4, pii, e09224.

Massey, T. H., Mercogliano, C. P., Yates, J., Sherratt, D. J., Lowe, J. 2006. Double-stranded DNA translocation: Structure and mechanism of hexameric FtsK. *Mol. Cell* 23, 457–469.

Moffitt, J. R., Chemla, Y. R., Aathavan, K., Grimes, S., Jardine, P. J., Anderson, D. L., Bustamante, C. 2009. Intersubunit coordination in a homomeric ring ATPase. *Nature* 457, 446–450.

Morita, M., Tasaka, M., Fujisawa, H. 1993. DNA packaging ATPase of bacteriophage T3. *Virology* 193, 748–752.

Oram, M., Sabanayagam, C., Black, L. W. 2008. Modulation of the packaging reaction of bacteriophage T4 terminase by DNA structure. *J. Mol. Biol.* 381, 61–72.

Orlova, E. V., Gowen, B., Droge, A., Stiege, A., Weise, F., Lurz, R., van, H. M., Tavares, P. 2003. Structure of a viral DNA gatekeeper at 10 A resolution by cryo-electron microscopy. *EMBO J.* 22, 1255–1262.

Saleh, O. A. et al. 2005. Analysis of DNA supercoil induction by FtsK indicates translocation without groove-tracking. *Nat Struct Mol Biol* 12:436–440.

Schwartz, C., De Donatis, G. M., Fang, H., Guo, P. 2013a. The ATPase of the phi29 DNA-packaging motor is a member of the hexameric AAA+ superfamily. *Virology* 443, 20–27.

Schwartz, C., De Donatis, G. M., Zhang, H., Fang, H., Guo, P. 2013b. Revolution rather than rotation of AAA+ hexameric phi29 nanomotor for viral dsDNA packaging without coiling. *Virology* 443, 28–39.

Serwer, P. 2010. A hypothesis for bacteriophage DNA packaging motors. *Viruses* 2, 1821–1843.

Simpson, A. A., Tao, Y., Leiman, P. G., Badasso, M. O., He, Y., Jardine, P. J., Olson, N. H. et al. 2000. Structure of the bacteriophage phi29 DNA packaging motor. *Nature* 408, 745–750.

Zhang, H., Schwartz, C., De Donatis, G. M., Guo, P. 2012. "Push through one-way valve" mechanism of viral DNA packaging. *Adv. Virus Res.* 83, 415–465.

Zhao, Z., Khisamutdinov, E., Schwartz, C., Guo, P. 2013. Mechanism of one-way traffic of hexameric phi29 DNA packaging motor with four electropositive relaying layers facilitating anti-parallel revolution. *ACS Nano* 7, 4082–4092.

chapter eight

Mechanism of rotary motors

8.1   Rotation motion in F_1

8.1.1   Single-molecule rotation assay of F_1

In F_oF_1 ATPase synthase, ATP synthesis from ADP and inorganic phosphate (P_i) is the basic reaction for cell energy production in animals, plants, and microorganisms. Cells have two major ATP production pathways. One is substrate-level ATP synthesis reaction like glycolysis, and the second is oxidative phosphorylation where electron transfer among electron carriers down the redox potential and drives respiratory chain proteins to pump protons for proton motive force (*pmf*) generation. ATP is synthesized upon the backflow of protons down *pmf*. F_oF_1 ATP synthase catalyzes this reaction as the terminal reaction of oxidative phosphorylation.

Dynamic behavior of the ATP-driven rotation of F_1 is well characterized in single-molecule rotation assay where the $\alpha_3\beta_3$ stator ring is immobilized on a coverslip and the rotation probe is attached onto the outwardly protruding part of the γ subunit of F_1 (Guo et al., 2016) (Figure 8.1a). In the presence of ATP, F_1 rotates the γ subunit counterclockwise as viewed from the F_o side. Since the unidirectional rotation of F_1 was first visualized from *Bacillus* PS3, several kinds of rotary ATPases—F_1's from *Escherichia coli* (Hisabori et al., 1999), chloroplast from spinach (Hisabori et al., 1999), human mitochondria (Suzuki et al., 2014), and V-ATPases from *Thermus thermophiles* (Imamura et al., 2003) and *Enterococcus hirae* (Minagawa et al., 2013)—were examined in the rotation assay. All rotary ATPases show the counterclockwise rotation without exception, implying a highly conserved structural basis for unidirectional rotation. Among them, F_1 from *Bacillus* PS3, termed TF_1 (thermophilic F_1), is the best characterized in terms of rotary dynamics. Therefore, the rotation features of F_1 introduced hereafter are based on the findings in the single-molecule rotation assay of TF_1 unless mentioned.

Rotation of the isolated F_1 motor driven by ATP hydrolysis was directly observed with an optical microscope (Noji et al., 1997; Okuno et al., 2011; Guo et al., 2016). Consistent with the pseudo threefold symmetry of the $\alpha_3\beta_3$ stator ring, F_1 rotates γ in discrete 120° steps (Yasuda et al., 1998), each coupled with a single turnover of ATP hydrolysis (Yasuda et al., 1998; Rondelez et al., 2005). An intermediate state was also observed

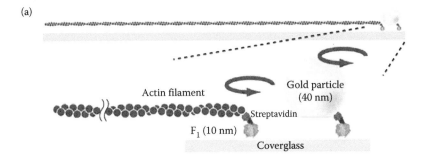

(a)

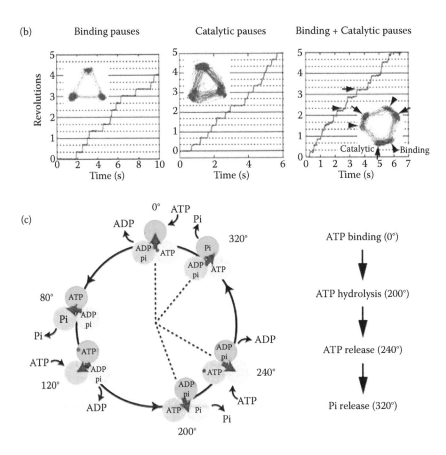

Figure 8.1 Single-molecule assay measuring the rotation of F_1. (a) Illustration of experimental setup. The $\alpha_3\beta_3$ ring is fixed on the glass surface. A probe (fluorescently labeled actin filament or 40-nm colloidal gold) is attached to the γ subunit. (b) (Left) Rotation of F_1 with three binding pauses separated by 120°, which is caused by slow ATP binding at 200 nM. Inset is the trajectory of the rotation.

(*Continued*)

Figure 8.1 (Continued) (Center) Rotation of a mutant F_1 (βE190D) with three catalytic pauses at 2-mM ATP. Each pause is caused by the extremely slow ATP hydrolysis by the mutant. (Right) Rotation of mutant F_1 (βE190D) at 2-μM ATP. Due to slow ATP binding and hydrolysis, six pauses are observed. The pauses before the 80° (arrowheads) and 40° (arrows) substeps correspond to binding and catalytic pauses, respectively. (c) Chemomechanical coupling scheme. Each circle indicates the chemical state of the catalytic sites. One catalytic site is highlighted in dark green. The central arrow (red) represents the angular position of the γ subunit. Each catalytic site retains the bound nucleotide as ATP until the γ subunit rotates 200° from the binding angle (0°). After a 200° rotation, the catalytic site executes the hydrolysis of ATP into ADP and Pi, each of which is released at 240° and 320°, respectively. (Adapted from Guo, P. et al., 2016. *Microbiol. Mol. Biol. Rev.* 80, 161–186, with permission.)

after ATP binding and before hydrolysis at an 80° angle from the ATP-waiting angle (Shimabukuro et al., 2003; Guo et al., 2016). Therefore, each 120° step can be resolved into 80° and 40° substeps (Figure 8.1b). Recent reaction scheme suggests that the 80° substep is driven by ATP binding and ADP release that occur on different β subunits, and that the 40° substep is initiated after hydrolysis of bound ATP and P_i release that also occur on different β subunits (Watanabe et al., 2010; Guo et al., 2016) (Figure 8.1c). It is under debate which reaction state the crystal structure of F_1 represents; although several lines of experiments suggest that most crystal structures of F_1 represent the intermediate state before the 40° substep (hydrolysis-waiting state), there are some crystal structures that are not explainable on this basis (Stracke et al., 2000; Menz et al., 2001; Guo et al., 2016).

8.1.2   Torque of F_1

Rotary torque that F_1 generates is estimated from the viscous friction against the rotating probe: Drag coefficient times angular velocity gives the torque that F_1 generates. Although some groups reported higher torque for F_1 from *E. coli* (EF$_1$) (Spetzler et al., 2006; Rees et al., 2012), most data show that F_1 generates torque of 40 pNnm irrespective of species (Yasuda et al., 1998; Noji et al., 1999; Panke et al., 2001). A theory of nonequilibrium physics, fluctuation theorem, was also employed as a new analytical method to estimate torque generation. This theorem also showed that F_1 generates torque of 40 pNnm (Hayashi et al., 2012). Considering that the torque is generated at the stator-rotor interface 1 nm from the rotation axis, the generated force is 40 pN, much larger than conventional linear motors like myosin and kinesin (Kinosita, Jr. et al., 1998). Interestingly, V-ATPase is known to generate slightly smaller torque than F_1, around 35 pN (Imamura et al., 2003). The magnitude of torque might reflect some physiological difference.

Torque times angular displacement gives the work done upon rotation. Therefore, the work done by F_1 per ATP is estimated to be 80 pNnm (40 pNnm $\times$ $2\pi/3$) (Yasuda et al., 1998). This value roughly agrees with the released free energy upon single turnover of ATP hydrolysis under physiological conditions, consistent with highly efficient energy conversion of F_oF_1 ATP synthase. Another important finding about torque generation is that F_1 exerts constant torque of 40 pNnm irrespective of viscous drag coefficient (Yasuda et al., 2001). This means that even when a large viscous marker significantly extends the timescale of mechanical rotation by 1000 times or more, F_1 bears high viscous drag force against a rotating probe, exerting constant torque. The rotation against large viscous drag is very similar to the situation under physiological conditions where the rotation torque of F_1 and F_o is mostly balanced, although the situation is not exactly the same from a thermodynamic point of view. If F_1 does not bear stall force condition and consumes ATP without rotation, F_oF_1 ATP synthase cannot interconvert *pmf* and free energy of ATP with high efficiency. Thus, robust torque generation against the high viscous friction is also relevant to the highly efficient energy conversion mechanism.

8.1.3   *Chemomechanical coupling of F_1*

To achieve high-energy conversion efficiency, high reversibility of chemomechanical coupling is required in addition to robust torque generation. The reversely coupled reaction of F_1, ATP production upon clockwise rotation was tested by forcibly rotating F_1 in a clockwise direction with magnetic tweezers. In the first experiment, a large number of F_1 molecules immobilized on a coverslip were simultaneously rotated and the synthesized ATP was detected with chemoluminescent reaction (Itoh et al., 2004). However, the exact number of active F_1 molecules was not clear so the coupling efficiency was not determined. In the following experiment, actively rotating F_1 molecules were individually encapsulated in a femtoliter chamber and rotated with magnetic tweezers (Rondelez et al., 2005). Due to the ultrasmall reaction volume, a small number of synthesized ATP molecules significantly increased ATP concentration in the chamber. After being released from forcible rotation, F_1 molecules resumed active ATP-hydrolyzing rotation that allowed for ATP measurement because the rotation velocity of F_1 is proportional to ATP concentration under the condition of low ATP concentration. This experiment showed that F_1 synthesizes ATP with high efficiency in the presence of the ε subunit. Although the role of the ε subunit has been extensively studied, it still remains elusive as to how the ε subunit enhances the coupling efficiency of ATP synthesis reaction of F_1.

As mentioned above, the step size of F_1 rotation is 120°, each of which is driven by a single turnover of ATP hydrolysis. However, it does not necessarily mean that each 120° step is driven by a particular β subunit. The actual reaction scheme of F_1 is much more complex. While all β subunits participate in torque generation for each 120° step, each β subunit is in a distinct conformational and thereby catalytic state (Abrahams et al., 1994). In addition, the 120° step is further resolved into 80° and 40° substeps, as shown by kinetic analysis. The 80° substep is initiated after ATP binding and ADP release. The 40° substep is initiated after hydrolysis of bound ATP and P_i release. The dwell angles before the 80° and 40° substeps are referred to as ATP-waiting angle and hydrolysis angle, respectively. To solve the puzzling reaction scheme, elaborated single-molecule experiments were conducted, such as simultaneous observation of fluorescently labeled ATP and rotation (Adachi et al., 2007), rotation assays with hybrid F_1 from the wild-type and mutant β subunits (Ariga et al., 2007), and single-molecule manipulation experiments (Watanabe et al., 2010). Most parts of the reaction scheme have been solved (Okuno et al., 2011). The recent reaction scheme (Figure 8.1c) is as follows: each β binds to ATP when the γ subunit is oriented to a particular ATP-waiting angle (defined as 0°). Then, β hydrolyzes bound ATP when the γ subunit rotates 200°. ADP inorganic phosphates (P_i) are released at 240° and 320°. The other two β subunits also undergo the same reaction pathway, while the phase of the catalytic state varies by +120° or −120°. Therefore, each 80° substep is initiated after ATP binding and ADP release that occurs at different β subunits, and the 40° substep is initiated after hydrolysis and Pi release that also occurs at different β subunits. It should be noted that the exact angle for P_i release is still under debate (Stracke et al., 2000; Shimo-Kon et al., 2010; Okazaki and Hummer, 2013).

Many experiments have showed that the large conformational change occurred in the catalytic β subunit (Abrahams et al., 1994; Masaike et al., 2008; Kobayashi et al., 2010). Upon nucleotide binding, β subunit undergoes the opened-to-closed conformational transition (Uchihashi et al., 2011), thus pushing the γ subunit as shown by NMR study (Kobayashi et al., 2010), single fluorophore imaging (Masaike et al., 2008), and AFM data, as expected from the crystal structure of MF_1. The elementary catalytic steps of ATP binding, hydrolysis, and release of ADP and P_i may serve as a major torque-generating step.

8.1.4 Torque generation steps of F_1

Large conformational transition does not necessarily guarantee large force generation. To estimate the contribution of ATP-binding and -hydrolysis steps for torque generation in F_1, single-molecule stall experiments were conducted where the equilibrium constants of ATP binding and hydrolysis were determined as a function of the rotary angle (Watanabe et al.,

2012). As a result, it was revealed that the equilibrium constant of ATP binding exponentially increases with the forward rotation of the γ subunit. This means that the F_1-ATP complex is progressively stabilized upon the rotation of γ, progressively releasing binding energy via γ rotation. This affinity change process is a typical induced fit. Although the hydrolysis process also showed the angle-dependent stabilization of the posthydrolysis state with the rotation of γ, the magnitude of equilibrium change was significantly small compared to the ATP-binding step. The estimated contribution of ATP binding for torque generation during the 80° substep is around 50% (up to 67%). The remaining torque generation is attributed to ADP release from a different β subunit. The contribution of hydrolysis in torque generation during the 40° substep is at most 21%, and P_i release is thought to be the major torque generation step in the 40° substep. Thus, it is evident that the role of chemical cleavage of phosphoester bond on the catalytic site in force generation is minor in F_1. The stall experiment also showed that the conformational transition upon the induced-fit process after ATP binding is one of the major torque generation steps. It is highly likely that the induced fit accompanies the opened-to-closed transition predicted from the crystal structure of MF_1. The torque generation upon ADP release would be induced-unfit, triggering closed-to-open transition. The conformational change upon P_i release is unclear. One feasible scenario is that the opening of the interface between the α and β subunits is coupled with P_i release (Ito et al., 2013).

As mentioned above, the induced-fit process following the first ATP docking to the catalytic site on the β subunit is a major torque-generating process of F_1. According to the crystal structure of MF_1, adenine ring is surrounded by aromatic amino acids while the phosphate moiety of ATP forms many hydrogen bonds with surrounding residues. Although crystal structures provide atomic pictures of conformations before and after conformational transition, it remained elusive as to which interactions are responsible in triggering the induced fit among adenine ring, ribose, and phosphoester of ATP. Recently, the role of adenine ring in torque generation was tested in rotation assay using a synthetic, base-free nucleotide (Arai et al., 2014). The impact of depletion of adenine ring was remarkable in the kinetics of rotation; the binding rate constant was decreased by six orders of magnitude. However, the synthetic base-free nucleotide supports powerful rotation of F_1 similarly to ATP; generated torque was not different between ATP-driven rotation and base-free nucleotide-driven rotation. Thus, it was proven that the interaction between adenine ring and the catalytic site is not responsible for torque generation. The principal role of adenine ring is rate enhancement of the first docking process, but it is not involved in the induced-fit process.

8.1.5   *Critical role of phosphate-binding sites in force generation*

The role of phosphoester in force generation was examined by the use of mutant F_1's, in which mutation is introduced at the phosphate-interacting residues. Targeted residues are p-loop lysine, glutamic acid of general base, and arginine finger, all of which are well known to be critical for retaining the catalytic power of F_1. Mutations at these residues lower the catalytic activity of F_1 down to an undetectable level in biochemical assay. p-Loop lysine, the lysine in the highly conserved phosphate-binding loop (p-loop with the common sequence GXXXXGKT/S) among NTPases, is well known to be the catalytically most critical residue in the p-loop; it is the common target to knock out ATPase activity of many ATPases. Glutamic acid of general base is also well conserved among ATPases that interact with the γ phosphate via a water molecule. Glutamic acid of general base has been thought to activate the intervening water molecule to induce nucleophilic attack of the water molecule to the γ phosphate (Guo et al., 2016). Recent quantum chemical calculation and single-molecule analysis revealed that the role of general base is not water activation but the enhancement of proton transfer from phosphoester to bulk solution (Hayashi et al., 2012). Arginine finger is also a highly conserved arginine residue among G proteins, AAA proteins, and RecA-type ATPases including F_1. Arginine finger is located at an interface of nucleotide-binding subunits (Guo et al., 2016). ATP binds the interface of two subunits, one of which possesses the majority of ATP-binding residues, while the other has little but catalytically critical residue that is arginine finger (Guo et al., 2016). The best-characterized arginine finger is that of G-protein activating protein (GAP). F_1 has arginine finger on the α subunit. The crystal structure of MF_1 with chemical analog of the γ phosphate suggests that arginine finger stabilizes the transient state of hydrolysis reaction, which is supported by biochemical (Nadanaciva et al., 1999), theoretical (Dittrich et al., 2003), and single-molecule studies (Komoriya et al., 2012).

The role of p-loop lysine, glutamic acid of general base, and arginine finger in torque generation was examined in the rotation assay with alanine mutants, in which alanine substitution was introduced at one of the target residues (Watanabe et al., 2014). Although all mutant F_1's support continuous and unidirectional rotation, the mutants showed significantly slower rotation than the wild type; the slowest was alanine mutant at p-loop lysine (1/1000 of the wild type). Thus, the role of mutations at phosphate-binding residues on kinetic power is significant as well as that of adenine ring. Prominent impact of the mutations was found in torque generation; all mutants were deficient for efficient torque generation. Alanine substitution at the general base and arginine finger halved torque. Mutations of p-loop lysine decreased torque down to 25% of the wild type. Thus,

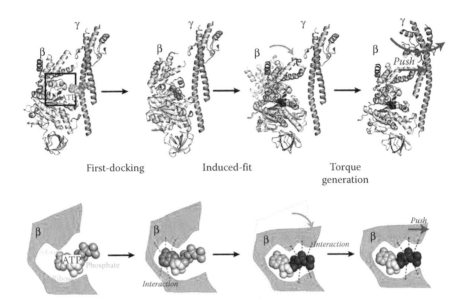

First-docking   Induced-fit   Torque
                                            generation

Figure 8.2 The opened-to-closed transition of the β subunit of MF$_1$. The accompanying swing motion of the C-terminal domain of the β subunit would push the γ subunit to induce the rotation. (Adapted from Guo, P. et al., 2016. *Microbiol. Mol. Biol. Rev.* 80, 161–186, with permission.)

phosphate-binding sites play a critical role in rate enhancement of both catalysis and torque generation.

Taking into account that the induced-fit process is a major torque-generating step, it was proposed that the progressive hydrogen bonds formation between the phosphate moiety of bound ATP and the catalytic residues are the main driving force of the induced-fit type conformational change; the opened-to-closed transition of the β subunit. The accompanying swing motion of the C-terminal domain of the β subunit would push the γ subunit to induce the rotation (Figure 8.2).

8.2   Rotation motion in F$_o$

Two hemichannel models of F$_o$: The rotation of the c-ring against the a_1b_2 complex of F$_o$ is driven by proton flow (Figure 3.1). Some bacteria have F$_o$ fueled by sodium motive force (Meier et al., 2005). As the whole structure of F$_o$ is not known, only a part of the ion-conducting pathway is revealed. Biochemical work has identified several charged residues in the transmembrane helices of the *a* and *c* subunits that would be directly involved in proton translocation. Among them, Asp or Glu of the *c* subunit resides in the middle of the C-terminal helix, and Arg of the *a* subunit, corresponding to

cAsp61 and aArg210 of *E. coli* F_o, are highly conserved among species and thought to play crucial roles in proton translocation. The crystal structure of the c-ring from *Ilyobacter tartaricus* F_o, a Na^+-transporting F_o, showed that the critical cGlu65 residues are occupied by Na^+ ions (Meier et al., 2005). The finding established that this conserved carboxyl residue is one of the ion-binding sites. However, other charged residues are not found in the c subunit in the vicinity of the carboxyl residue, suggesting that the a subunit has other parts of the ion-conducting pathway. The most widely accepted model on proton translocation in F_o is the two-channel model, originally proposed by W. Junge (Junge et al., 1997). This model assumes that the a subunit possesses two hemichannels, each of which spans half of the membrane but toward different sides. The hemichannels connect the ion-conducting carboxyl residue of the c subunit (Figure 3.1). Each channel is in contact with a different c subunit, which are adjacent to each other. Thus, the a subunit interacts with two c subunits, each contacting via a different half channel of the a subunit. The proposed mechanism of proton transfer in the ATP synthesis mode is as follows: A proton enters the half channel exposed to the periplasmic side (or intermembrane space of the mitochondria) and is then transferred to the carboxyl residue of the c subunit. This protonation induces the transformation of the carboxyl residue to an ion-locked conformation (Pogoryelov et al., 2010) (Figure 3.1) (Pogoryelov et al., 2009) and neutralizes the negative charge of the residue, allowing the c subunit to rotate apart from the a subunit toward the hydrophobic lipid layer. At the same time, the neighboring c subunit on the counterclockwise side returns from the lipid layer to form contacts with the other half channel, which has a hydrophilic environment. This contact promotes deprotonation of the carboxyl residue and induces the transformation of the c subunit to an ion-unlocked form (Symersky et al., 2012). The released proton then enters into the cytoplasmic space. Thus, proton translocation accompanies the rotation of the c-ring. For each proton, the c-ring makes one turn. Importantly, the proton translation pathway shows intrinsic chirality for the unidirectional rotation of F_o; individual protons are translated after one clockwise turn (in ATP synthesis mode) as represented by the cyan arrow in Figure 3.1. This is highly consistent with the concept of "helical proton channel" (Minamino et al., 2008). The role of the conserved Arg in the a subunit is thought to prevent the proton shortcut without c-ring rotation (Mitome et al., 2010). In the ATP-driven proton-pumping mode, this sequence of events is reversed.

The characterization of c-ring rotation against the ab_2 complex in F_o is an important clue to elucidating the working principle of the F_o motor. Compared with the rotary dynamics of F_1, F_o has been less characterized due to challenges in purifying the whole complex of F_oF_1 without losing any subunits (some detergents are known to promote subunit dissociation) and handling the enzyme with a highly hydrophobic part. Thus far, detergent-solubilized, fully functional F_oF_1 was subjected to the rotation assay under ATP-hydrolysis conditions (Sambongi et al., 1999; Ueno et al., 2005; Ishmukhametov et al., 2010). The rotation of *E. coli* F_oF_1 (EF_oF_1) in reconstituted nanodiscs was recently reported. In these measurements, torque-generating F_1 dragged the *c*-ring in a counterclockwise direction. Thereby, unless the *c*-ring becomes a kinetic bottleneck, the dynamic features of *c*-ring rotation are not investigable. Only a few studies reported intervening pauses representing the multiple potential bumps that were attributed to transiently formed interaction between *c*-ring and the ab_2 complex (Ishmukhametov et al., 2010).

References

Abrahams, J. P., Leslie, A. G., Lutter, R., Walker, J. E. 1994. Structure at 2.8 A resolution of F1-ATPase from bovine heart mitochondria. *Nature* 370, 621–628.

Adachi, K., Oiwa, K., Nishizaka, T., Furuike, S., Noji, H., Itoh, H., Yoshida, M., Kinosita, K. 2007. Coupling of rotation and catalysis in F-1-ATPase revealed by single-molecule imaging and manipulation. *Cell* 130, 309–321.

Arai, H. C., Yukawa, A., Iwatate, R. J., Kamiya, M., Watanabe, R., Urano, Y., Noji, H. 2014. Torque generation mechanism of F-1-ATPase upon NTP binding. *Biophy. J.* 107, 156–164.

Ariga, T., Muneyuki, E., Yoshida, M. 2007. F1-ATPase rotates by an asymmetric, sequential mechanism using all three catalytic subunits. *Nat. Struct. Mol. Biol.* 14, 841–846 (Erratum in *Nat. Struct. Mol. Biol.* 14, 984).

Dittrich, M., Hayashi, S., Schulten, K. 2003. On the mechanism of ATP hydrolysis in F(1)-ATPase. *Biophys. J.* 85, 2253–2266.

Doering, C., Ermentrout, B., Oster, G. 1998. Rotary DNA motors. *Biophys. J.* 69, 2256–2267.

Guo, P., Noji, H., Yengo, C. M., Zhao, Z., Grainge, I. 2016. Biological nanomotors with revolution, linear, or rotation motion mechanism. *Microbiol. Mol. Biol. Rev.* 80, 161–186.

Hayashi, S., Ueno, H., Shaikh, A. R., Umemura, M., Kamiya, M., Ito, Y., Ikeguchi, M., Komoriya, Y., Iino, R., Noji, H. 2012. Molecular mechanism of ATP hydrolysis in F1-ATPase revealed by molecular simulations and single-molecule observations. *J. Am. Chem. Soc.* 134, 8447–8454.

Hisabori, T., Kondoh, A., Yoshida, M. 1999. The gamma subunit in chloroplast F-1-ATPase can rotate in a unidirectional and counter-clockwise manner. *FEBS Lett.* 463, 35–38.

Imamura, H., Nakano, M., Noji, H., Muneyuki, E., Ohkuma, S., Yoshida, M., Yokoyama, K. 2003. Evidence for rotation of V-1-ATPase. *Proc. Natl. Acad. Sci. USA* 100, 2312–2315.

Ishmukhametov, R., Hornung, T., Spetzler, D., Frasch, W. D. 2010. Direct observation of stepped proteolipid ring rotation in *E. coli* FoF1-ATP synthase. *EMBO J.* 29, 3911–3923.

Ito, Y., Yoshidome, T., Matubayasi, N., Kinoshita, M., Ikeguchi, M. 2013. Molecular dynamics simulations of yeast F-1-ATPase before and after 16 degrees rotation of the gamma subunit. *J. Phys. Chem. B* 117, 3298–3307.

Itoh, H., Takahashi, A., Adachi, K., Noji, H., Yasuda, R., Yoshida, M., Kinosita, K. 2004. Mechanically driven ATP synthesis by F-1-ATPase. *Nature* 427, 465–468.

Junge, W., Lill, H., Engelbrecht, S. 1997. ATP synthase: An electrochemical transducer with rotatory mechanics. *Trends Biochem. Sci.* 22, 420–423.

Kinosita, K., Jr., Yasuda, R., Noji, H., Ishiwata, S., Yoshida, M. 1998. F1-ATPase: A rotary motor made of a single molecule. *Cell* 93, 21–24.

Kobayashi, M., Akutsu, H., Suzuki, T., Yoshida, M., Yagi, H. 2010. Analysis of the open and closed conformations of the beta subunits in thermophilic F-1-ATPase by solution NMR. *J. Mol. Biol.* 398, 189–199.

Komoriya, Y., Ariga, T., Iino, R., Imamura, H., Okuno, D., Noji, H. 2012. Principal role of the arginine finger in rotary catalysis of F-1-ATPase. *J. Biol. Chem.* 287, 15134–15142.

Masaike, T., Koyama-Horibe, F., Oiwa, K., Yoshida, M., Nishizaka, T. 2008. Cooperative three-step motions in catalytic subunits of F(1)-ATPase correlate with 80 degrees and 40 degrees substep rotations. *Nat. Struct. Mol. Biol.* 15, 1326–1333.

Meier, T., Polzer, P., Diederichs, K., Welte, W., Dimroth, P. 2005. Structure of the rotor ring of F-type Na+-ATPase from *Ilyobacter tartaricus*. *Science* 308, 659–662.

Menz, R. I., Walker, J. E., Leslie, A. G. 2001. Structure of bovine mitochondrial F(1)-ATPase with nucleotide bound to all three catalytic sites: Implications for the mechanism of rotary catalysis. *Cell* 106, 331–341.

Minagawa, Y., Ueno, H., Hara, M., Ishizuka-Katsura, Y., Ohsawa, N., Terada, T., Shirouzu, M. et al. 2013. Basic properties of rotary dynamics of the molecular motor *Enterococcus hirae* V-1-ATPase. *J. Biol. Chem.* 288, 32700–32707.

Minamino, T., Imada, K., Namba, K. 2008. Molecular motors of the bacterial flagella. *Curr. Opin. Struct. Biol.* 18, 693–701.

Mitome, N., Ono, S., Sato, H., Suzuki, T., Sone, N., Yoshida, M. 2010. Essential arginine residue of the F-0-a subunit in F0F1-ATP synthase has a role to prevent the proton shortcut without c-ring rotation in the F-0 proton channel. *Biochem. J.* 430, 171–177.

Nadanaciva, S., Weber, J., Wilke-Mounts, S., Senior, A. E. 1999. Importance of F-1-ATPase residue alpha-Arg-376 for catalytic transition state stabilization. *Biochemistry* 38, 15493–15499.

Noji, H., Hasler, K., Junge, W., Kinosita, K., Jr., Yoshida, M., Engelbrecht, S., 1999. Rotation of *Escherichia coli* F(1)-ATPase. *Biochem. Biophys. Res. Commun.* 260, 597–599.

Noji, H., Yasuda, R., Yoshida, M., Kinosita, K., Jr. 1997. Direct observation of the rotation of F1-ATPase. *Nature* 386, 299–302.

Okazaki, K., Hummer, G. 2013. Phosphate release coupled to rotary motion of F-1-ATPase. *Proc. Natl. Acad. Sci. USA* 110, 16468–16473.

Okuno, D., Iino, R., Noji, H. 2011. Rotation and structure of FoF1-ATP synthase. *J. Biochem.* 149, 655–664.

Panke, O., Cherepanov, D. A., Gumbiowski, K., Engelbrecht, S., Junge, W. 2001. Viscoelastic dynamics of actin filaments coupled to rotary F-ATPase: Angular torque profile of the enzyme. *Biophys. J.* 81, 1220–1233.

Pogoryelov, D., Krah, A., Langer, J. D., Yildiz, O., Faraldo-Gomez, J. D., Meier, T. 2010. Microscopic rotary mechanism of ion translocation in the F-0 complex of ATP synthases. *Nat. Chem. Biol.* 6, 891–899.

Pogoryelov, D., Yildiz, O., Faraldo-Gomez, J. D., Meier, T. 2009. High-resolution structure of the rotor ring of a proton-dependent ATP synthase. *Nat. Struct. Mol. Biol.* 16, 1068–1073.

Rees, D. M., Montgomery, M. G., Leslie, A. G. W., Walker, J. E. 2012. Structural evidence of a new catalytic intermediate in the pathway of ATP hydrolysis by F-1-ATPase from bovine heart mitochondria. *Proc. Natl. Acad. Sci. USA* 109, 11139–11143.

Rondelez, Y., Tresset, G., Nakashima, T., Kato-Yamada, Y., Fujita, H., Takeuchi, S., Noji, H. 2005. Highly coupled ATP synthesis by F1-ATPase single molecules. *Nature* 433, 773–777.

Sambongi, Y., Iko, Y., Tanabe, M., Omote, H., Iwamoto-Kihara, A., Ueda, I., Yanagida, T., Wada, Y., Futai, M. 1999. Mechanical rotation of the c subunit oligomer in ATP synthase (F0F1): Direct observation. *Science* 286, 1722–1724.

Shimabukuro, K., Yasuda, R., Muneyuki, E., Hara, K. Y., Kinosita, K., Yoshida, M. 2003. Catalysis and rotation of F-1 motor: Cleavage of ATP at the catalytic site occurs in 1 ms before 40 degrees substep rotation. *Proc. Natl. Acad. Sci. USA* 100, 14731–14736.

Shimo-Kon, R., Muneyuki, E., Sakai, H., Adachi, K., Yoshida, M., Kinosita, K. 2010. Chemo-mechanical coupling in F-1-ATPase revealed by catalytic site occupancy during catalysis. *Biophys. J.* 98, 1227–1236.

Spetzler, D., York, J., Daniel, D., Fromme, R., Lowry, D., Frasch, W. 2006. Microsecond time scale rotation measurements of single F-1-ATPase molecules. *Biochemistry* 45, 3117–3124.

Stracke, J. O., Fosang, A. J., Last, K., Mercuri, F. A., Pendas, A. M., Llano, E., Perris, R., Di Cesare, P. E., Murphy, G., Knauper, V. 2000. Matrix metalloproteinases 19 and 20 cleave aggrecan and cartilage oligomeric matrix protein (COMP)1. *FEBS Lett.* 478, 52–56.

Suzuki, T., Tanaka, K., Wakabayashi, C., Saita, E., Yoshida, M. 2014. Chemomechanical coupling of human mitochondrial F1-ATPase motor. *Nat. Chem. Biol.* 10(11), 930–936.

Symersky, J., Pagadala, V., Osowski, D., Krah, A., Meier, T., Faraldo-Gomez, J. D., Mueller, D. M. 2012. Structure of the c(10) ring of the yeast mitochondrial ATP synthase in the open conformation. *Nat. Struct. Mol. Biol.* 19, 485–491.

Uchihashi, T., Iino, R., Ando, T., Noji, H. 2011. High-speed atomic force microscopy reveals rotary catalysis of rotorless F-1-ATPase. *Science* 333, 755–758.

Ueno, H., Suzuki, T., Kinosita, K., Yoshida, M. 2005. ATP-driven stepwise rotation of FOF1-ATP synthase. *Proc. Natl. Acad. Sci. USA* 102, 1333–1338.

Watanabe, R., Iino, R., Noji, H. 2010. Phosphate release in F-1-ATPase catalytic cycle follows ADP release. *Nat. Chem. Biol.* 6, 814–820.

Watanabe, R., Matsukage, Y., Yukawa, A., Tabata, K. V., Noji, H. 2014. Robustness of the rotary catalysis mechanism of F-1-ATPase. *J. Biol. Chem.* 289, 19331–19340.

Watanabe, R., Okuno, D., Sakakihara, S., Shimabukuro, K., Iino, R., Yoshida, M., Noji, H. 2012. Mechanical modulation of catalytic power on F-1-ATPase. *Nat. Chem. Biol.* 8, 86–92.

Yasuda, R., Noji, H., Kinosita, K., Jr., Yoshida, M. 1998. F1-ATPase is a highly efficient molecular motor that rotates with discrete 120 degree steps. *Cell* 93, 1117–1124.

Yasuda, R., Noji, H., Yoshida, M., Kinosita, K., Jr., Itoh, H. 2001. Resolution of distinct rotational substeps by submillisecond kinetic analysis of F1-ATPase. *Nature* 410, 898–904.

chapter nine

Mechanism of linear motors

Linear motors are characterized by their ability to move or generate force through interaction with a specific cellular track (cytoskeletal filaments, DNA, or RNA). The motion of many of these motors has been described as "walking," where they advance one foot—or head, in this case—at a time. Two cytoskeletal motors, myosin and kinesin, are members of the P-loop NTPases that share structural homology in the nucleotide binding region (Kull et al., 1998). Dynein is a more complex motor that has six ATP binding regions and is a member of the AAA+ ATPase family (Roberts et al., 2013). To highlight the features of linear motors, in this section we will focus on the myosin superfamily, consisting of around 2000 motors divided into 35 classes that are ubiquitously expressed in eukaryotic cells (Odronitz & Kollmar, 2007). Myosins are capable of interacting cyclically, with actin filaments utilizing the chemical energy derived from ATP hydrolysis to perform mechanical work (Sellers & Goodson, 1995). By virtue of this mechanical work, these motors can translocate actin filaments or act as tethers/anchors to generate tension and force. Moreover, certain classes of myosin motors can also act as point-to-point transporters, thus individually moving cargo on actin filaments.

A common feature among linear motors is that small conformational changes in the nucleotide binding region are transmitted to the effector (actin or microtubule) binding regions and to the force generating elements of the motor which are capable of large conformational changes that drive motion (Vale & Milligan, 2000). This two-way allosteric communication originating from the nucleotide binding region ensures that the motor maintains contact with the track while generating force. In myosins, the lever arm consists of the light chain binding region. Variability in the length of the light chain binding region in different myosin isoforms helped to prove the hypothesis that this region functions as a lever arm (Tyska & Warshaw, 2002). In kinesin, the neck linker or coiled-coil stalk has been demonstrated to be critical for movement (Rice et al., 1999; Vale & Milligan, 2000). In dynein, the coiled-coil of the stalk, which connects the AAA ring and the microtubule binding domain, is known to change conformation in a nucleotide-dependent manner and function as a force generating element (Roberts et al., 2013).

The energy derived from ATP binding and hydrolysis is efficiently utilized for mechanical motion of linear motors (Woledge et al., 1985; Cooke, 1997; Howard, 2001). Futile ATPase cycles are preventable because the release of the products of ATP hydrolysis is several orders of magnitude slower in the absence of the cytoskeletal track (Malnasi-Csizmadi & Kovacs, 2010). In myosin, there is strong coupling between the mechanical and chemical cycles, since actin binding is proposed to accelerate the structural changes in the force generating element. A key difference between kinesin and myosin is that ATP binding does not detach kinesin from the microtubule but is associated with the conformational change in the force generating element (Kull & Endow, 2013). A major difference between the three families of linear motors (myosin, kinesin, and dynein) lies in the type of surface on which they move. Kinesin and dynein move along microtubules and thus can participate in long-range transport while myosin can move along actin filaments or produce force by moving actin filaments when anchored in the membrane or contractile apparatus. In addition, kinesin and dynein move in opposite directions, with dynein heading toward the center of the cell and most kinesins from the center of the cell outward. All three linear motors have demonstrated that they can walk in a hand-over-hand fashion, where they advance one step at a time in a unidirectional fashion, not unlike walking using their heads as feet (Yildiz & Selvin, 2005). ATP binding and hydrolysis causes motors to cycle between weak and strong binding states along the filament, with ATP binding and turnover providing the energy to detach from one binding site and attach to a new one by taking forward steps.

9.1 Conserved catalytic cycle of myosins

The conserved myosin ATPase cycle first described by Lymn–Taylor demonstrates the main features of the actin-activated ATPase cycle (Lymn & Taylor, 1971). The myosin motor domain is an ATPase that is strongly activated upon binding to actin. In the absence of any nucleotide, myosin binds to actin tightly and forms a rigor complex. ATP binding to myosin causes myosin to detach from actin and enter the weak binding states. During the detached states, ATP is hydrolyzed and the lever arm region of myosin primes itself into a pre-power stroke state (Recovery Stroke, Figure 9.1) (Agafonov et al., 2009; Nesmelov et al., 2011; Muretta et al., 2013). Thereafter, myosin complexed with the hydrolysis products rebinds actin in a weak binding state. This is followed by the release of P_i first and then ADP, which is stimulated by the binding of the complex to actin (actin-activated product release). During actin-activated phosphate release, myosin pulls on the actin filament performing mechanical work which is produced by the swing of the lever arm (power stroke, Figure 9.1) (Malnasi-Csizmadia & Kovacs, 2010; Muretta et al., 2015; Trivedi et al., 2015; Tyska & Warshaw, 2002). An additional power stroke has been shown to occur in some myosins during the ADP release step, which is thought to be associated with strain sensitivity (Veigel et al., 1999; Nyitrai & Geeves, 2004)

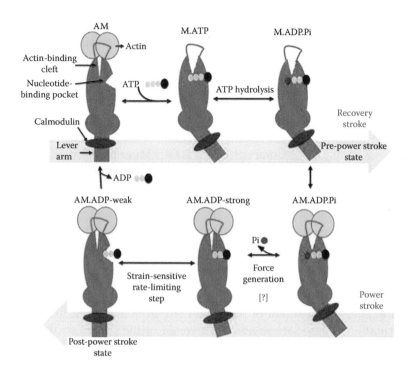

Figure 9.1 A simplified schematic representation of the ATPase cycle of myosin. The proposed mechanism of how the motor is primed (recovery stroke) and generates force by coupling movement of the lever to key steps in the ATPase cycle. (Adapted from Trivedi, D. V. 2014. Allosteric communication and force generation in myosin motors. Doctoral dissertation. Pennsylvania State University, University Park, PA; Guo, P. et al. 2016. *Microbiol. Mol. Biol. Rev.* 80, 161–186, with permission.)

or the ability of these mechanoenzymes to adapt to different loads. A two-ADP-state model, one with strong ADP binding affinity and the other with weak affinity, has been proposed based on kinetic, structural, and mechanical studies (Rosenfeld et al., 2005; Hannemann et al., 2005; Oguchi et al., 2008; Trivedi et al., 2013).

9.2 Force generation in myosins

The long alpha-helix emanating from the motor domain of myosin motors is termed the lever arm. The reversible movement of the lever arm as a function of actin binding or ATP binding is what drives the force generating mechanism of myosin motors. The movement of the lever arm in an actin-bound state is thought to drive the majority of force generation and is thus called the power stroke. Following the power stroke, the lever arm can be returned to the pre-power stroke state by a conformational change referred to as the recovery stroke. The recovery stroke occurs in an ATP bound and actin-unbound state. The lever arm movement during the recovery and power stroke stages of the catalytic cycle has been probed indirectly by a number of studies (Trivedi, 2014; Guo et al., 2016). Monitoring the intrinsic fluorescence signal originating from the tryptophan residue located at the distal end of the relay-helix near the converter region has yielded insights into the coupling of the open–closed transition of switch II active site and apparent position of the lever arm (Malnasi-Csizmadia et al., 2000; Conibear et al., 2004). More recently, there have been studies measuring the lever arm swing in *Dictyostelium* myosin by utilizing strategically placed FRET probes on the relay-helix (Muretta et al., 2013; Agofonov, 2009; Nesmelov, 2011). In this work, they indirectly measured the swing by correlating the kinked to straight conformation of the relay-helix to the power stroke. This study definitively shows that by utilizing EPR and transient time resolved FRET, helix straightening occurs after actin binding and before P_i release. The authors hypothesize that relay-helix straightening gates P_i release, which in turn provides the thermodynamic driving force for force generation. The group also reported that the reverse movement of the relay-helix from a straight to a kinked conformation is associated with the reversal of the power stroke or the recovery stroke. The straight to kinked transition of the relay-helix occurs after ATP binding and before hydrolysis. Hence, these studies report movement of the lever arm during the recovery stroke and power stroke based on the conformation of the relay-helix.

Modeling studies based on structural models of *Dictyostelium* myosin II have yielded insights into the structural mechanism of the recovery stroke (Fischer et al., 2005; Koppole et al., 2007). However, owing to a lack of crystal structures in the actin-bound states, it has been difficult to perform modeling studies of the movement of the lever arm during the power stroke. Preller et al. (Preller & Holmes, 2013) also performed targeted molecular dynamics simulations with *Dictyostelium* Myo II and found that soon after actin binding, a 16° rotation of the L50 kDa domain puts strain on a helix that is connected to the actin-binding site. The strain twists the beta sheet connected to this helix, which can drive the power stroke without opening switch I or switch II. They propose that during the power stroke, switch II moves, thus opening an exit route for P_i to escape, which would explain actin-activated phosphate release.

Several studies based on muscle fiber mechanics have given insights into the timing of the force generating step in the intact sarcomere. Dantzig et al. (Dantzig et al., 1992) measured force generation and decline in tension after photolysis of caged P_i on glycerol-extracted fibers from rabbit psoas muscle. In the tension recordings, soon after P_i release, a lag of several milliseconds occurs before the force declines. The authors propose a two-step mechanism of force generation and P_i release with force generation preceding the release of P_i from the active site. Another report investigated the timing of P_i binding/release and the mechanism of force generation in rabbit fast-twitch muscle fibers by employing the method of sinusoidal analysis (Kawai & Halvorson, 1991). These studies propose that a conformational isomerization precedes P_i release. Further, they also infer a distinct ADP bound state of the crossbridge and propose the transition between the two ADP bound states as a rate-limiting step of the cycle. Other studies performed on rabbit psoas muscles by Nagano & Yanagida (1984), Millar & Homsher (1990), and Sleep et al. (2005), and on frog skeletal muscle fibers by Brozovich et al. (1988) also show similar results that are consistent with a rapid power stroke before P_i release. Alternatively, laser temperature jump experiments performed on rabbit psoas muscle fibers predict a mechanism wherein P_i release provides the energy to generate tension, by swinging the lever arm in a force generating state (Davis & Rodgers, 1995).

Single molecule studies have demonstrated a reversibility of the force generating lever arm swing under high loads without the net utilization of ATP (Takagi et al., 2006). Sellers et al. (Sellers & Veigel, 2010) report a reversibility of the power stroke with myosin V at intermediate loads (2–5 pN). A recent study by Capitanio et al. (2012) demonstrated that a decrease in amplitude of the working stroke in muscle fibers at high loads is due to a premature dissociation pathway that becomes more populated at higher forces. The temporal resolution of the working stroke obtained by this group was within an interval of 2 ms after initial binding of skeletal muscle myosin to the actin filament. This provides evidence of a fast power stroke that may precede P_i release, especially in skeletal myosin where P_i release is relatively slow compared to other steps in the catalytic cycle. An additional swing of the lever arm is reported to occur in a number of myosins, which is associated with the actomyosin-ADP state. Uemura et al. (2004) demonstrate that the working stroke of myosin V is composed of two substeps. They predict that the ADP associated swing occurs around the time after the release of P_i and before formation of the weak-ADP binding state. Another study proposes a model wherein a 5 nm substep of the power stroke is accomplished by a dimeric myosin V followed by the release of ADP (Veigel et al., 2002). This sub-step acts as a gate to relieve the strain that is generated by the binding of both heads to the actin filament. Biphasic working strokes are also known to occur in Brush Border Myosin I (Veigel et al., 1999), rat liver myosin I (Veigel et al., 1999), and smooth muscle myosin (Veigel et al., 2003).

The most recent work on the precise timing of lever arm movement in relation to the catalytic cycle of myosins comes from the Yengo (Trivedi, 2014; Trivedi et al., 2015), Thomas (Muretta et al., 2015), and Houdusse/Sweeney (Llinas et al., 2015; Houdusse & Sweeney, 2016) laboratories wherein they examined the movement of the lever-arm during power stroke and recovery stroke stages. The Yengo (Trivedi et al., 2015) and Thomas (Muretta et al., 2015) labs utilized strategically placed FRET probes to directly measure the movement of the lever arm by steady-state and transient time-resolved FRET during the catalytic cycle of myosin. This work settled a long-standing controversy regarding the precise timing of the power stroke in relation to the product release steps. They demonstrated that after binding to actin, the first power stroke occurs before Pi release, and a second stroke occurs in the actomyosin.ADP states. They also demonstrated that the myosin undergoes the recovery stroke immediately after ATP binding, and hydrolysis of ATP locks the lever arm into the pre-power stroke state. However, on the basis of crystal structures, the Houdusse laboratory (Llinas et al., 2015) inferred that the power stroke occurs after release of Pi. The steady-state and transient time resolved FRET techniques provide information on the conformational changes of the lever arm as the myosin is undergoing its catalytic cycle, while crystal structures provide detailed information on one particular snapshot in the catalytic cycle. A model proposing to resolve this discrepancy has also been proposed (Houdusse & Sweeney, 2016).

Overall, the key steps in the mechanochemical cycle of myosin are as follows. In the first step, the myosin head, with no nucleotide bound to its active site, is bound to the actin filament. ATP binding dramatically reduces the affinity of myosin for actin, and simultaneously primes the lever arm for force generation by undergoing the recovery stroke. ATP is then hydrolyzed, and the myosin head contains ADP and Pi bound to its active site. The head then makes contact with actin which triggers a rapid power stroke followed by Pi release and a subsequent slower power stroke prior to ADP release. Once the active site is empty, it is ready to rebind ATP and repeat the force generation cycle.

As expected from their structural similarity, kinesin and myosin share some common features in their structural mechanism of motion generation (Kull & Endow, 2013). The conserved nucleotide binding region of kinesin can be directly linked to the microtubule binding and force generating elements. The switch II region is linked to a long helix (alpha-4) that essentially functions as the relay helix connecting the microtubule and nucleotide binding regions. The relay helix also impacts the conformation of the neck linker by altering interactions with an adjacent helix (alpha-6). The conformation of the switch I region also plays a role in microtubule binding by coordination of key microtubule binding elements (alpha-3 and loop-8).

Allosteric communication pathways are at the heart of force generation in the linear motors. It is critically important to uncover these coupling pathways, which will help us understand the molecular basis of disease-causing mutations in these motors and may facilitate drug design paradigms to target these diseases. Moreover, understanding the basis behind the force generation mechanism of these motors can also help us design artificial motors that can be utilized for nanotechnological applications.

References

Agafonov, R. V., Negrashov, I. V., Tkachev, Y. V., Blakely, S. E., Titus, M. A., Thomas, D. D., Nesmelov, Y. E. 2009. Structural dynamics of the myosin relay helix by time-resolved EPR and FRET. *Proc. Natl. Acad. Sci. USA* 106, 21625–21630.

Brozovich, F. V., Yates, L. D., Gordon, A. M. 1988. Muscle force and stiffness during activation and relaxation. Implications for the actomyosin ATPase. *J. Gen. Physiol.* 91, 399–420.

Capitanio, M., Canepari, M., Maffei, M., Beneventi, D., Monico, C., Vanzi, F., Bottinelli, R., Pavone, F. S. 2012. Ultrafast force-clamp spectroscopy of single molecules reveals load dependence of myosin working stroke. *Nat. Methods* 9, 1013–1019.

Conibear, P. B., Malnasi-Csizmadia, A., Bagshaw, C. R. 2004. The effect of F-actin on the relay helix position of myosin II, as revealed by tryptophan fluorescence, and its implications for mechanochemical coupling. *Biochemistry* 43, 15404–15417.

Cooke, R. 1997. Actomyosin interaction in striated muscle. *Physiol. Rev.* 77, 671–697.

Dantzig, J. A., Goldman, Y. E., Millar, N. C., Lacktis, J., Homsher, E. 1992. Reversal of the cross-bridge force-generating transition by photogeneration of phosphate in rabbit psoas muscle fibres. *J. Physiol.* 451, 247–278.

Davis, J. S., Rodgers, M. E. 1995. Force generation and temperature-jump and length-jump tension transients in muscle fibers. *Biophys. J.* 68, 2032–2040.

Fischer, S., Windshugel, B., Horak, D., Holmes, K. C., Smith, J. C. 2005. Structural mechanism of the recovery stroke in the myosin molecular motor. *Proc. Natl. Acad. Sci. USA* 102, 6873–6878.

Guo, P., Noji, H., Yengo, C. M., Zhao, Z., Grainge, I. 2016. Biological nanomotors with revolution, linear, or rotation motion mechanism. *Microbiol. Mol. Biol. Rev.* 80, 161–186.

Hannemann, D. E., Cao, W., Olivares, A. O., Robblee, J. P., De la Cruz, E. M. 2005. Magnesium, ADP, and actin binding linkage of myosin V: Evidence for multiple myosin V-ADP and actomyosin V-ADP states. *Biochemistry* 44, 8826–8840.

Houdusse, A., Sweeney, H. L. 2016. How myosin generates force on actin filaments. *Trends Biochem. Sci.* 41, 989–997.

Howard, J. 2001. *Mechanics of Motor Proteins and the Cytoskeleton*. Sinauer, Sunderland, Massachusetts.

Kawai, M., Halvorson, H. R. 1991. Two step mechanism of phosphate release and the mechanism of force generation in chemically skinned fibers of rabbit psoas muscle. *Biophys. J.* 59, 329–342.

Koppole, S., Smith, J. C., Fischer, S. 2007. The structural coupling between ATPase activation and recovery stroke in the myosin II motor. *Structure* 15, 825–837.

Kull, F. J., Endow, S. A. 2013. Force generation by kinesin and myosin cytoskeletal motor proteins. *J. Cell. Sci.* 126, 9–19.

Kull, F. J., Vale, R. D., Fletterick, R. J. 1998. The case for a common ancestor: Kinesin and myosin motor proteins and G proteins. *J. Muscle. Res. Cell. Motil.* 19, 877–886.

Llinas, P., Isabet, T., Song, L., Ropars, V., Zong, B., Benisty, H., Sirigu, S., Morris, C., Kikuti, C., Safer, D., Sweeney, H. L., Houdusse, A. 2015. How actin initiates the motor activity of myosin. *Dev. Cell.* 33, 401–412.

Lymn, R. W., Taylor, E. W. 1971. Mechanism of adenosine triphosphate hydrolysis by actomyosin. *Biochemistry* 10, 4617–4624.

Malnasi-Csizmadia, A., Kovacs, M. 2010. Emerging complex pathways of the actomyosin power stroke. *Trends Biochem. Sci.* 35, 684–690.

Malnasi-Csizmadia, A., Woolley, R. J., Bagshaw, C. R. 2000. Resolution of conformational states of *Dictyostelium* myosin II motor domain using tryptophan (W501) mutants: Implications for the open-closed transition identified by crystallography. *Biochemistry* 39, 16135–16146.

Millar, N. C., Homsher, E. 1990. The effect of phosphate and calcium on force generation in glycerinated rabbit skeletal muscle fibers. A steady-state and transient kinetic study. *J. Biol. Chem.* 265, 20234–20240.

Muretta, J. M., Petersen, K. J., Thomas, D. D. 2013. Direct real-time detection of the actin-activated power stroke within the myosin catalytic domain. *Proc. Natl. Acad. Sci. USA* 110, 7211–7216.

Muretta, J. M., Rohde, J. A., Johnsrud, D. O., Cornea, S., Thomas, D. D. 2015. Direct real-time detection of the structural and biochemical events in the myosin power stroke. *Proc. Natl. Acad. Sci. USA* 112, 14272–14277.

Nagano, H., Yanagida, T. 1984. Predominant attached state of myosin cross-bridges during contraction and relaxation at low ionic strength. *J. Mol. Biol.* 177, 769–785.

Nesmelov, Y. E., Agafonov, R. V., Negrashov, I. V., Blakely, S. E., Titus, M. A., Thomas, D. D. 2011. Structural kinetics of myosin by transient time-resolved FRET. *Proc. Natl. Acad. Sci. USA* 108, 1891–1896.

Nyitrai, M., Geeves, M. A. 2004. Adenosine diphosphate and strain sensitivity in myosin motors. *Philos. Trans. R. Soc. Lond. B. Biol. Sci.* 359(1452), 1867–1877.

Odronitz, F., Kollmar, M. 2007. Drawing the tree of eukaryotic life based on the analysis of 2,269 manually annotated myosins from 328 species. *Genome. Biol.* 8, R196.

Oguchi, Y., Mikhailenko, S. V., Ohki, T., Olivares, A. O., De La Cruz, E. M., Ishiwata, S. 2008. Load-dependent ADP binding to myosins V and VI: Implications for subunit coordination and function. *Proc. Natl. Acad. Sci. USA*, 105(22), 7714–7719.

Preller, M., Holmes, K. C. 2013. The myosin start-of-power stroke state and how actin binding drives the power stroke. *Cytoskeleton (Hoboken.)* 70, 651–660.

Rice, S. et al. 1999. Structural change in the kinesin motor protein that drives motility. *Nature* 402, 778–784.

Roberts, A. J., Kon, T., Knight, P. J., Sutoh, K., Burgess, S. A. 2013. Functions and mechanics of dynein motor proteins. *Nat. Rev. Mol. Cell. Biol.* 14, 713–726.

Rosenfeld, S. S., Houdusse, A., Lee Sweeney, H. 2005. Magnesium regulates ADP dissociation from myosin V. *J. Biol. Chem.* 280, 6072–6079.

Sellers, J. R., Goodson, H. V. 1995. Motor proteins 2: Myosin. *Protein Profile* 2, 1323–1423.

Sellers, J. R., Veigel, C. 2010. Direct observation of the myosin-Va power stroke and its reversal. *Nat. Struct. Mol. Biol.* 17, 590–595.

Sleep, J., Irving, M., Burton, K. 2005. The ATP hydrolysis and phosphate release steps control the time course of force development in rabbit skeletal muscle. *J. Physiol.* 563, 671–687.

Takagi, Y., Homsher, E. E., Goldman, Y. E., Shuman, H. 2006. Force generation in single conventional actomyosin complexes under high dynamic load. *Biophys. J.* 90, 1295–1307.

Trivedi, D. V. 2014. Allosteric communication and force generation in myosin motors. *Doctoral dissertation*. Pennsylvania State University, University Park, PA.

Trivedi, D. V., Muretta, J. M., Swenson, A. M., Thomas, D. D., Yengo, C. M. 2013. Magnesium impacts myosin V motor activity by altering key conformational changes in the mechanochemical cycle. *Biochemistry* 52(27), 4710–4722.

Trivedi, D. V., Muretta, J. M., Swenson, A. M., Davis, J. P., Thomas, D. D., Yengo, C. M. 2015. Direct measurements of the coordination of lever arm swing and the catalytic cycle in myosin V. *Proc. Natl. Acad. Sci. USA* 112, 14593–14598.

Tyska, M. J., Warshaw, D. M. 2002. The myosin power stroke. *Cell. Motil. Cytoskeleton* 51, 1–15.

Uemura, S., Higuchi, H., Olivares, A. O., De la Cruz, E. M., Ishiwata, S. 2004. Mechanochemical coupling of two substeps in a single myosin V motor. *Nat. Struct. Mol. Biol.* 11, 877–883.

Vale, R. D., Milligan, R. A. 2000. The way things move: Looking under the hood of molecular motor proteins. *Science.* 288, 88–95.

Veigel, C., Coluccio, L. M., Jontes, J. D., Sparrow, J. C., Milligan, R. A., Molloy, J. E. 1999. The motor protein myosin-I produces its working stroke in two steps. *Nature* 398, 530–533.

Veigel, C., Molloy, J. E., Schmitz, S., Kendrick-Jones, J. 2003. Load-dependent kinetics of force production by smooth muscle myosin measured with optical tweezers. *Nat. Cell. Biol.* 5, 980–986.

Veigel, C., Wang, F., Bartoo, M. L., Sellers, J. R., Molloy, J. E. 2002. The gated gait of the processive molecular motor, myosin V. *Nat. Cell. Biol.* 4, 59–65.

Woledge, R. C., Curtin, N. A., Homsher, E. 1985. Energetic aspects of muscle contraction. *Monogr. Physiol. Soc.* 41, 1–357.

Yildiz, A., Selvin, P. R. 2005. Fluorescence imaging with one nanometer accuracy: Application to molecular motors. *Acc. Chem. Res.* 38, 574–582.

chapter ten

Applications of the revolving motor

10.1 The discovery of a new method for the development of highly potent drugs (Pi et al., 2016)

As discussed previously, the key step of genome packaging is usually accomplished by a biomotor using ATP in viral reproduction, including but not limited to dsDNA/dsRNA bacteriophages (Zhao et al., 2013), adenoviruses (Ostapchuk and Hearing, 2005), poxviruses (Chan et al., 2009), human cytomegaloviruses (HCMV) (Hwang and Bogner, 2002), and herpes simplex viruses (HSV) (Roos et al., 2007). Intrigued by the unique viral structure and the packaging mechanisms, extensive studies have been carried out to elucidate the fundamentals in protein/DNA, protein/RNA and RNA/DNA interactions in the quest for new prototypes of biological machines or new antiviral drugs. Several decades ago, Jonathon King pointed out that studies on viral DNA packaging will lead to the discovery of new antiviral drugs (King and Casjens, 1974). This idea has stimulated scientists to pursue further study on viral DNA packaging mechanism to uncover better drug targets. Indeed, our study on phi29 packaging motor revealed that, several key motor components composed of multimeric complexes, such as hexameric pRNA ring, hexameric gp16 ATPase, and ATP with more than 10,000 copies to package one genome (Schwartz and Guo, 2013; Shu et al., 2015), may serve as better drug targets than a monomeric genomic DNA. Comparing the viral packaging inhibition efficiency of mock drugs targeting each machine with different stoichiometry, we found that the inhibition efficiency increased exponentially to the stoichiometry of the targeted biocomplex (Figure 10.1a,b) (Pi et al., 2015, 2016; Shu et al., 2015).

Based on the new revolving mechanism of biomotor introduced in the earlier chapters of this book, a method for developing highly efficient inhibitory drugs was described by targeting biological machines with high stoichiometry and exhibiting sequential action mechanism (Pi et al., 2015, 2016; Shu et al., 2015). As for the drug inhibition efficiency on its target biological entity, a determining factor is the ratio of the drugged ($T_{inactive}$) to undrugged (T_{active}) target components. Within each viral infected cell, the higher percentage of drugged machine ($T_{inactive}$) leads to higher inhibitory efficiency. The percentage of drugged machine can be calculated from the binomial distribution or Yang Hui's Triangle (Figure 10.1c):

$$(p+q)^Z = \binom{Z}{0}p^Z + \binom{Z}{1}p^{Z-1}q^1 + \dots \binom{Z}{Z-1}p^1q^{Z-1} + \binom{Z}{Z}q^Z = \sum_{M=0}^{Z}\binom{Z}{M}p^Mq^{Z-M} \tag{10.1}$$

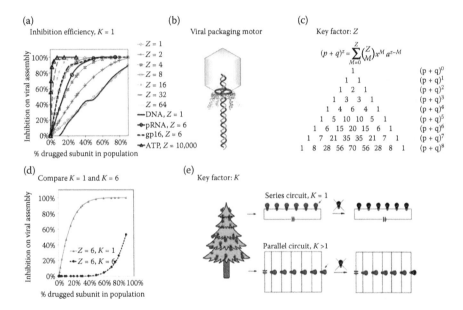

Figure 10.1 Drug inhibition efficiency is correlated with the stoichiometry of the targeted biocomplex. (a) Targeting 10,000 subunits of ATP showed the strongest inhibition efficiency, compared to those of the 6-subunit pRNA or gp16 and the 1-subunit genomic DNA. (b) Illustration of the phi29 viral DNA-packaging motor. (c) One key factor regarding drug potency is the stoichiometry of the homomeric complex serving as a target: the ratio of the functional complex can be calculated from Yang Hui's triangle. (d) Comparison of drug inhibition efficiencies in a complex with a $Z = 6$ and a $K = 1$ and with a $Z = 6$ and a $K = 6$. Here, K is the number of blocked subunits required to inhibit the function of the whole complex target. (e) Illustration showing that K is a key factor for drug potency. When $K = 1$ as in the ATPase sequential motion, the system is similar to a series circuit of Christmas lights; one broken bulb will turn off the whole chain. When $K > 1$, the chain will not be completely turned off until all bulbs inside the parallel circuit are broken. (Adapted from Pi, F. et al. 2016. *J. Virol.* 90, 8036–8046, with permission.)

Here, p and q represent the fraction of drugged inactive and non-drugged active subunits in the population, Z is the stoichiometry of the target, and M represents drugged subunits in each biocomplex.

1. The first intrinsic factor of the target for potent drug development is Z, which is the stoichiometry of the homomeric complex serving as drug target. The ratio of functional complex changes dramatically with Z. In Equation 10.1, assuming that the ratio of drugged inactive subunits (p) and undrugged active subunits (q) in the population are fixed, when the target machine contains only one subunit, the ratio equals to q as derived from $Z = 1$ in binomial distribution (Equation 10.2). Here $p + q = 100\%$.

$$(p+q)^1 = p+q \tag{10.2}$$

However, when the homomeric target complex contains multiple subunits, a binomial distribution formula with a higher order (Equation 10.3) is applied. When $Z = 4$ then,

$$(p+q)^4 = p^4 + 4p^3q^1 + 6p^2q^2 + 4p^1q^3 + q^4 \tag{10.3}$$

In this case, the probability of target machine complex possessing four copies of inactive subunit is p^4; three copies of inactive and one copy of active subunit is $4p^3q$; two copies of inactive and two copies of uninhibited subunits is $6p^2q^2$; three copies of active subunits is $4pq^3$; and, four copies of active subunit is q^4. Assuming that 70% (p) of subunits are inactivated by drugs, the percentage of active machines containing four copies of active subunits is q^4, which is $(0.3) 4 = 0.8\%$. However, when $Z = 1$ and $p = 70$, the ratio of the uninhibited portion equals to 30%. Thus, the stoichiometry (Z) of targeted machine has significant contribution to the ratio of survival rate, which directly correlates with the drug inhibition efficiency on viral replication.

2. The second intrinsic factor of the target for potent drug development is K, which is the number of drugged subunits required to block the function of the complex. Stoichiometry only has a multiplicative effect on inhibition efficiency when $K = 1$. Reinterpreting this statement with an example of a homo-hexamer machine ($Z = 6$) as drug target, and 70% (p) of subunits are inactivated by drugs: if $K = 1$, the ratio of active target complex will be $0.3^6 = 0.07\%$; while $K = 6$ the ratio will be $1 - 0.7^6 = 88.2\%$. $K = 1$ showed much stronger inhibition compared to the case of $K = 6$, where the active complex ratio was 88.2% at the same blocking subunit ratio (Figure 10.1d). $K = 1$ is a key factor for multisubunit complexes as potent drug targets.

3. The third intrinsic factor is that the homomeric complex acts through a sequential or coordination mechanism. Sequential or coordination action means that each subunit of the complex works in turn to complete the function of the complex (Figure 10.1e). Blocking any step of the sequential action results in deactivation of the complex. That meets the definition of $K = 1$ in a homomeric complex. Analogous to a string of Christmas lights, where one broken light bulb will turn off the entire chain, one inhibited subunit will deactivate the entire complex and consequently their biological activity.

With the above three intrinsic factors, we find that viral DNA packaging motors are ideal targets for high efficient antiviral drugs, since these motors are machines with high stoichiometry and their subunits work sequentially. Here, "stoichiometry" differs from a conventional concept used in drug development. Conventionally, stoichiometry refers to the number of drug molecules bound to each virion or target related to viral replication. In this case, stoichiometry refers to the number of identical subunits in the viral machine as target for drug development (Pi et al., 2015, 2016; Shu et al., 2015).

Traditionally, it was almost impossible to prove this concept by comparing efficacies of two drugs acting on two targets with different stoichiometries. It is very challenging to demonstrate whether the difference of drug efficiency is contributed by the essentiality of the two different targets in viral function, or the affinity of drug molecules in target binding, or the stoichiometry of the targeted machine. The phi29 DNA packaging system offered an excellent model to prove this concept. It has the sensitivity of 10^9 PFU for *in vitro* assays using purified components (Lee and Guo, 1994), which made it possible to use inactive mutant components to represent the drugged substrate. Therefore, the ratio of the drugged and undrugged subunits can be explicitly defined by binomial distribution calculation (Pi et al., 2015, 2016; Shu et al., 2015).

The concept described here for high efficient viral inhibition may have an impact in antiviral therapy by introducing dominant negative proteins (Lee et al., 2014) or inactive mutant proteins into the cell, either by intracellular expression using viral vectors for gene delivery or direct introduction of proteins into cell (Trottier et al., 1996; Chen et al., 1997; Fang et al., 2014; Shu et al., 2015). This involves the incorporation of mutant protein subunits into a multimeric complex identified as drug target. If a multimeric complex is identified with high stoichiometry and $K = 1$ due to the sequential action of the homomeric subunits, then incorporating one mutant subunit into the complex would inactivate the complex completely. Since the viral machine is composed of Z subunits, one drugged subunit per complex would only work when the intracellular drug concentration is high. But if the strategy is to apply a dominant negative protein, such as the dominant negative phospholamban in cardiac gene therapy (Lee et al., 2014), then more augmented effect of the mutant protein subunits will be expected as the Z value increases.

Besides the viral DNA packaging motors, viral machines with high stoichiometry and operated by cooperative sequential action with $Z > 1$ and $K = 1$ are ubiquitous. These viral machines are involved in many aspects in the viral life cycle. These machines include, but are not limited to, viral DNA polymerase, viral RNA polymerase, reverse transcriptase, chaperons, viral genome repair enzymes, viral integrase, membrane pores for viral DNA or RNA trafficking, viral motors for dsRNA packaging, as well as other machines involved in viral entry, motion, and trafficking. Drugs targeting to these viral machines will be highly efficient (Pi et al., 2015, 2016; Shu et al., 2015). Design of potent viral drugs with high specificity is also possible. For example, if a specific high stoichiometry machine is identified in a virus, specific drugs targeting subunits of this machine will be highly efficient.

Since viral motors share certain common structures and operation mechanisms, methods of targeting homomeric multisubunit complexes should have general applications in antiviral drug discovery. Homomeric channel proteins, such as the homotetramer M2 proton channel protein, have been shown to be a better target for anti-influenza drugs (Wang et al., 1993). Amantadine and rimantadine inhibit the influenza virus through this mechanism by entering the barrel of the tetrameric M2 channel and blocking proton translocation function (Figure 10.2a) (Philippe et al., 2013). RNA-dependent RNA-polymerase NS5B, which plays an important role in hepatitis C virus (HCV) replication,

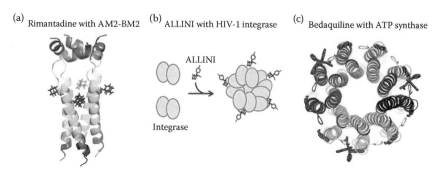

(a) Rimantadine with AM2-BM2   (b) ALLINI with HIV-1 integrase   (c) Bedaquiline with ATP synthase

ALLINI

Integrase

Figure 10.2 Drugs that target viral motors with high stoichiometry. (a) Homotetrameric AM2-BM2 channel protein of influenza virus complexed with the inhibitory drug molecule rimantadine (PDB accession number 2RLF); (b) illustration of allosteric HIV-1 integrase inhibitors that act by promoting high-order oligomerization of integrase to block its activity; and (c) bedaquiline that targets ATP synthase (PDB accession number 4V1H). (Adapted from Pi, F. et al. 2016. *J. Virol.* 90, 8036–8046, with permission.)

exists as a homomeric oligomer. NS5B–NS5B intermolecular interaction is essential for both initiation and elongation of RNA synthesis (Lopez-Jimenez et al., 2014). Targeting the oligomeric protein NS5B could also be helpful to delineate new and powerful antiviral strategies (Pi et al., 2015, 2016; Shu et al., 2015).

The method for discovery of potent dsDNA viral therapeutics can be extended to other viruses. For example, allosteric HIV-1 integrase inhibitors (ALLINIs) that potently impair HIV-1 replication in cell culture are currently undergoing clinical trials (Garmann et al., 2015; Pettersen et al., 2004). While HIV-1 integrase functions as a tetramer to catalyze covalent insertion of the viral cDNA into human chromosome, ALLINIs bind at the IN dimer interface and promote cooperative, higher order oligomerization, resulting in inactive protein (Figure 10.2b). Such an innovative mode of action is energetically much more favorable than the competitive mechanism of action of preventing interaction between subunits and could potentially pave a way for developing novel compounds for other viral targets utilizing multiprotein subunits (Pi et al., 2015, 2016; Shu et al., 2015).

The method described here has general applications in other biological systems. In fact, the first drug approved to treat multidrug-resistant tuberculosis, bedaquiline (Lakshmanan and Xavier, 2013), acts on ATP synthase which is a multisubunit biomotor (Figure 10.2c) (Watanabe et al., 2014; Yasuda et al., 1998; Kinosita et al., 1998; Stock et al., 1999; Boyer, 1999; Adachi et al., 2000; Hara et al., 2000; Masaike et al., 2000; Wada et al., 2000; Okazaki and Hummer, 2013; Ito et al., 2013; Arai et al., 2014). Although this drug's inventors were not aware of the concept of targeting multisubunit complexes, the success of this drug supports the notion of using the multisubunit complex as a potent drug target. Cancer or bacterial mutant multisubunit ATPase can also be used as targets. Drug developers can simply check the published literature and find the multisubunit machine as a drug target. For cancer treatment, the key is to find multisubunit machines with mutations (Pi et al., 2016).

References

Adachi, K., Yasuda, R., Noji, H., Itoh, H., Harada, Y., Yoshida, M., Kinosita, K., Jr. 2000. Stepping rotation of F1-ATPase visualized through angle-resolved single-fluorophore imaging. *Proc. Natl. Acad. Sci. USA* 97, 7243–7247.

Arai, H. C., Yukawa, A., Iwatate, R. J., Kamiya, M., Watanabe, R., Urano, Y., Noji, H. 2014. Torque generation mechanism of F-1-ATPase upon NTP binding. *Biophys. J.* 107, 156–164.

Boyer, P. D. 1999. What makes ATP synthase spin? *Nature* 402, 247–249.

Chan, K. W., Yang, C. H., Lin, J. W., Wang, H. C., Lin, F. Y., Kuo, S. T., Wong, M. L., Hsu, W. L. 2009. Phylogenetic analysis of parapoxviruses and the C-terminal heterogeneity of viral ATPase proteins. *Gene* 432, 44–53.

Chen, C., Trottier, M., Guo, P. 1997. New approaches to stoichiometry determination and mechanism investigation on RNA involved in intermediate reactions. *Nucleic Acids Symp. Ser.* 36, 190–193.

Fang, H., Zhang, P., Huang, L. P., Zhao, Z., Pi, F., Montemagno, C., Guo, P. 2014. Binomial distribution for quantification of protein subunits in biological nanoassemblies and functional nanomachines. *Nanomedicine.* 10, 1433–1440.

Garmann, R. F., Gopal, A., Athavale, S. S., Knobler, C. M., Gelbart, W. M., Harvey, S. C. 2015. Visualizing the global secondary structure of a viral RNA genome with cryo-electron microscopy. *RNA* 21, 877–886.

Hara, K. Y., Noji, H., Bald, D., Yasuda, R., Kinosita, K., Jr., Yoshida, M. 2000. The role of the DELSEED motif of the beta subunit in rotation of F1-ATPase. *J. Biol. Chem.* 275, 14260–14263.

Hwang, J. S., Bogner, E. 2002. ATPase activity of the terminase subunit pUL56 of human cytomegalovirus. *J. Biol. Chem* 277, 6943–6948.

Ito, Y., Yoshidome, T., Matubayasi, N., Kinoshita, M., Ikeguchi, M. 2013. Molecular Dynamics Simulations of Yeast F-1-ATPase before and after 16 degrees Rotation of the gamma Subunit. *J. Phys. Chem. B* 117, 3298–3307.

King, J., Casjens, S. 1974. Catalytic head assembly protein in virus morphogenesis. *Nature* 251, 112–119.

Kinosita, K., Jr., Yasuda, R., Noji, H., Ishiwata, S., Yoshida, M. 1998. F1-ATPase: A rotary motor made of a single molecule. *Cell* 93, 21–24.

Lakshmanan, M., Xavier, A. S. 2013. Bedaquiline - The first ATP synthase inhibitor against multi drug resistant tuberculosis. *J. Young. Pharm.* 5, 112–115.

Lee, C. S., Guo, P. 1994. A highly sensitive system for the assay of *in vitro* viral assembly of bacteriophage phi29 of *Bacillus subtilis*. *Virology* 202, 1039–1042.

Lee, J. Y., Finkelstein, I. J., Arciszewska, L. K., Sherratt, D. J., Greene, E. C. 2014. Single-molecule imaging of FtsK translocation reveals mechanistic features of protein-protein collisions on DNA. *Mol. Cell* 54, 832–843.

Lopez-Jimenez, A. J., Clemente-Casares, P., Sabariegos, R., Llanos-Valero, M., Bellon-Echeverria, I., Encinar, J. A., Kaushik-Basu, N., Froeyen, M., Mas, A. 2014. Hepatitis C virus polymerase-polymerase contact interface: Significance for virus replication and antiviral design. *Antiviral Res.* 108, 14–24.

Masaike, T., Mitome, N., Noji, H., Muneyuki, E., Yasuda, R., Kinosita, K., Yoshida, M. 2000. Rotation of F(1)-ATPase and the hinge residues of the beta subunit. *J. Exp. Biol.* 203(Pt 1), 1–8.

Okazaki, K., Hummer, G. 2013. Phosphate release coupled to rotary motion of F-1-ATPase. *Proc. Natl. Acad. Sci. USA* 110, 16468–16473.

Ostapchuk, P., Hearing, P. 2005. Control of adenovirus packaging. *J. Cell. Biochem.* 96, 25–35.

Pettersen, E. F., Goddard, T. D., Huang, C. C., Couch, G. S., Greenblatt, D. M., Meng, E. C., Ferrin, T. E. 2004. UCSF chimera - A visualization system for exploratory research and analysis. *J. Comput. Chem.* 25, 1605–1612.

Philippe, N., Legendre, M., Doutre, G., Coute, Y., Poirot, O., Lescot, M., Arslan, D. et al. 2013. Pandoraviruses: Amoeba viruses with genomes up to 2.5 Mb reaching that of parasitic eukaryotes. *Science* 341, 281–286.

Pi, F., Vieweger, M., Zhao, Z., Wang, S., Guo, P. 2015. Discovery of a new method for potent drug development using power function of stoichiometry of homomeric biocomplexes or biological nanomotors. *Expert Opin. Drug Deliv.* 13, 23–36.

Pi, F., Zhao, Z., Chelikani, V., Yoder, K., Kvaratskhelia, M., Guo, P. 2016. Development of potent antiviral drugs inspired by viral hexameric DNA-packaging motors with revolving mechanism. *J. Virol.* 90, 8036–8046.

Roos, W. H., Ivanovska, I. L., Evilevitch, A., Wuite, G. J. L. 2007. Viral capsids: Mechanical characteristics, genome packaging and delivery mechanisms. *Cellular and Molecular Life Sciences* 64, 1484–1497.

Schwartz, C., Guo, P. 2013. Ultrastable pRNA hexameric ring gearing hexameric phi29 DNA-packaging motor by revolving without rotating and coiling. *Curr. Opin. Biotechnol.* 24(4), 581–590.

Shu, D., Pi, F., Wang, C., Zhang, P., Guo, P. 2015. New approach to develop ultra-high inhibitory drug using the power-function of the stoichiometry of the targeted nanomachine or biocomplex. *Nanomedicine* 10, 1881–1897.

Stock, D., Leslie, A. G., Walker, J. E. 1999. Molecular architecture of the rotary motor in ATP synthase. *Science* 286, 1700–1705.

Trottier, M., Zhang, C. L., Guo, P. 1996. Complete inhibition of virion assembly *in vivo* with mutant pRNA essential for phage phi29 DNA packaging. *J. Virol.* 70, 55–61.

Wada, Y., Sambongi, Y., Futai, M. 2000. Biological nano motor, ATP synthase F(o) F(1): From catalysis to gammaepsilonc(10-12) subunit assembly rotation. *Biochim. Biophys. Acta* 1459, 499–505.

Wang, C., Takeuchi, K., Pinto, L. H., Lamb, R. A. 1993. Ion channel activity of influenza A virus M2 protein: Characterization of the amantadine block. *J. Virol.* 67, 5585–5594.

Watanabe, R., Matsukage, Y., Yukawa, A., Tabata, K. V., Noji, H. 2014. Robustness of the rotary catalysis mechanism of F-1-ATPase. *J. Biol. Chem.* 289, 19331–19340.

Yasuda, R., Noji, H., Kinosita, K., Jr., Yoshida, M. 1998. F1-ATPase is a highly efficient molecular motor that rotates with discrete 120 degree steps. *Cell* 93, 1117–1124.

Zhao, H., Christensen, T. E., Kamau, Y. N., Tang, L. 2013. Structures of the phage Sf6 large terminase provide new insights into DNA translocation and cleavage. *Proc. Natl. Acad. Sci. USA* 110, 8075–8080.

Concluding remarks and perspectives

Nanobiomotors are tiny machines that utilize a primary energy source in mechanical work. They are crucial to the sustenance of living systems, since they provide for the majority of forms of biological motion, helping to direct cellular components in proper destinations, package DNA, contract muscles, and perform a variety of other functions. Biomotors exhibit a diversity of complex structures. Most have the same basic components, including a mechanical frame (usually composed of proteins) with both moving and static parts, powered by an energy supply. This energy is typically derived from the hydrolysis of ATP, which leads to conformational changes in the motor protein and results in movement, but other motors use energy produced from ion gradients. These motors are typically divided into categories based on the type of motion displayed: The most well-studied motors are categorized tentatively into linear motors, rotary motors, and revolving motors. The action of revolving that enables a motor free of coiling and torque has solved many puzzles that have occurred throughout the history of viral DNA packaging motor studies. It also helps to clarify questions concerning the structure, stoichiometry, and functioning of DNA translocation motors. This book uses bacteriophages phi29, HK97, SPP1, P22, T4, and T7 as well as bacterial DNA translocase FtsK and SpoIIIE as examples to elucidate the structure, function, and mechanisms. Some motors described here use a hexameric ATPase to revolve around the dsDNA sequentially. ATP binding induces conformational change and possibly an entropy alteration in ATPase to a high affinity toward dsDNA; but ATP hydrolysis triggers another entropic and conformational change in ATPase to a low affinity for DNA, by which dsDNA is pushed toward an adjacent ATPase subunit. The rotation and revolving mechanisms can be distinguished by the size of the channel: the channels of rotation motors are equal to or smaller than 2 nm, whereas channels of revolving motors are larger than 3 nm. Rotation motors use parallel threads to operate a right-handed channel, while revolving motors use a left-handed channel to drive the right-handed DNA in an antiparallel arrangement. Coordination of several vector factors in the same direction makes viral DNA-packaging motors unusually powerful and effective. The revolving mechanism avoids DNA coiling in translocating the lengthy genomic dsDNA helix.

Competing interests

Peixuan Guo's Sylvan G. Frank Endowed Chair position in Pharmaceutics and Drug Delivery is funded by the CM Chen Foundation. PG is a consultant of Oxford Nanopore and Nanobio Delivery Pharmaceutical Co., Ltd. He is the cofounder of P&Z Biological Technology LLC.

Glossary

DNA (deoxyribonucleic acid) versus RNA (ribonucleic acid)—both are nucleic acids that carry genetic information, but there are differences between them:

1. Structure
   a. Where DNA contains deoxyribose, RNA contains ribose. Ribose has one more -OH unit than deoxyribose, and deoxyribose has -H on the second carbon in the ring.
   b. With few exceptions, RNA is single-stranded; DNA is double-stranded.
   c. The chains of nucleotides tend to be shorter with strands of RNA.
2. Stability
   a. DNA is more stable, whereas RNA is more reactive.
3. Formation
   a. Where DNA is protected and lasts longer, RNA is constantly made, deconstructed, and reused.
   b. DNA is self-replicating; RNA is synthesized from DNA.
4. Functions
   a. DNA is responsible for the long-term storage of genetic information, which it transmits to form new cells and organisms. Its function is to transmit genetic information from one generation to the next.
   b. RNA plays a role in the formation of proteins by transferring genetic code from the cell nucleus to ribosomes. It also directs protein synthesis. In some organisms, RNA is used to transmit genetic information.
5. Location
   a. DNA tends to be found in the nucleus and mitochondria.
   b. RNA is usually found in the cytoplasm but can also be in the cell nucleus or ribosome.

5′ and 3′, with reference to DNA and RNA—′ stands for "prime." Each carbon atom in a ribose (for RNA) or deoxyribose (for DNA) molecule is numbered along the ring with a ′ after each one. To form a chain of molecules, one deoxyribose attaches to another with the 5′ carbon of one attaching to the 3′ carbon of another one using a phosphate group. Once they form a chain, the deoxyribose on one end has an unattached 5′ carbon, and the other end has an unattached 3′ carbon. In the case of double-stranded DNA, the other strand will be the opposite (where there was a 5′ carbon one the end of one strand, the other ends with a 3′ carbon and vice-versa).

μm   Micrometer or micron, measuring one millionth of a meter. There are 1000 nanometers in a micrometer.

Å   Angstrom. It measures one ten-billionth of a meter or one tenth of a nanometer.

Actin   A protein in muscle cells that, along with myosin, plays an important role in muscle contraction and relaxation. It is also involved in many other processes, including cell movement. The term comes from the Latin *āctus*, which means "motion."

Adenosine diphosphate (ADP)   A nucleotide composed of adenosine and two linked phosphate groups that is converted to ATP for the storage of energy in living cells. Intermediate between ATP and AMP.

Adenosine monophosphate (AMP)   A nucleotide composed of adenosine and one acidic phosphate group that is reversibly convertible to ADP and ATP.

Antichiral   A property referring to a molecule that is superimposable with its mirror image. Because both molecules are, for all intents and purposes, identical, they cannot be left- or right-handed as in the case of chiral molecules.

Arginine   A basic, essential amino acid that occurs naturally in proteins and is used in the biosynthesis of proteins. Plays many important roles in the human body including cell division, decreased healing and repair time, and removing ammonia from the body.

Arginine finger   An arginine-rich structural motif that has a number of different functions in ATPases, including serving as a bridge between two adjacent subunits.

Additional strand, conserved E (ASCE)   A family of p-loop NTPases containing Walker A and B motifs.

Assay   A procedure for measuring the concentration, activity, and effect of change to a substance by testing the organism and then comparing this measurement to a standard.

Adenosine triphosphate (ATP)   A nucleotide that transports chemical energy within cells for metabolism but has many other functions.

ATP is the main energy source for most cellular functions, including the synthesis of DNA and RNA.

ATPase An enzyme that catalyzes the formation of ATP from ADP, inorganic phosphate, and energy, thus releasing energy for use in another biochemical reaction. Converts ADP to ATP. Hydrolyzes ATP to ADP and phosphate. There are different forms of ATPase (F-ATPase [F_1-ATPase and F_0-ATPase], V-ATPase, etc.)

ATP binding The act of forming a chemical bond with ATP.

ATP hydrolysis The process by which cells break down ATP to use its energy. Water molecules split apart ATP to form ADP, and energy is released.

ATP synthase Creates the cellular energy force ATP.

Bacterial flagellum Helical-shaped, whip-like appendage that allows bacteria to move through liquid medium.

Bacterial binary fission Process of bacterial cell division that causes it to reproduce; mechanics and sequence are different than mitosis.

Bacteriophage Virus that infects and replicates within bacteria, consisting of a protein capsid enclosing a single piece of DNA. In Greek, *-phage* "to devour." Also known simply as a "phage."

Binding site An area on the surface of a protein or piece of DNA or RNA where other molecules may bind or interact. Biomotors move from binding site to binding site.

Biomotor A type of nanomachine normally composed of proteins that does mechanical work on a cellular level. Cells contain a complex network of tracks inhabited by biomotors which move along these tracks. They are responsible for a myriad of cellular processes and are the agents of movement in living things.

bp Base pair. A pair of hydrogen bonds that connect complementary strands of DNA or RNA. In DNA, adenine forms a base pair with thymine; but in RNA, adenine forms a base pair with uracil rather than thymine. In both DNA and RNA, guanine pairs with cytosine.

Calmodulin (stands for calcium-modulated protein; abbreviated as CaM) A messenger protein in eukaryotic cells that senses calcium levels in the body and relays signals to control various processes.

Capsid The protein shell of a virus that encloses its genetic material.

Chirality A property referring to a molecule that cannot be superimposed on its mirror image due to geometric differences between them. Most biological molecules are chiral. Pairs of chiral molecules can be either right- or left-handed. For example, your hands are chiral – there is no way you could rotate them and have them look identical to one another. Even though they are mirror images of one another, the thumbs are on the opposite side when you hold them both face-up. In the same way, a molecule can look identical when faced with its mirror image because the arrangement of the atoms appears to be the same, when you put them side-by-side, they are not in the same location in both molecules. Comes from the Greek *cheir,* meaning "hand."

Cleavage   The act of severing the sugar-phosphate bonds between nucleotides in DNA.

Closed circular DNA (or cccDNA)   A double-stranded DNA structure that forms a loop and therefore has no free ends.

Concatemer   A long strand of DNA containing multiple copies of the same DNA arranged end-to-end.

Confocal microscopy (or confocal laser scanning microscopy)   A form of microscopy that allows for a more controlled and limited depth of focus than conventional microscopy, increases optical resolution by eliminating out-of-focus light, and allows for the reconstruction of three-dimensional structures by collecting sets of images at different depths within a thick object.

Conformation   The spatial relationship of atoms in a molecule and the infinite number of ways that they can be arranged or rotated.

Conformational change   Change in the shape of a macromolecule.

Cryo-electron microscopy (Cryo-EM)   Electron microscopy where the sample is studied at cryogenic temperatures, without dyes or fixatives, allowing scientists to study minute objects at molecular resolution. Reveals the hidden machinery of a cell and is used in single-particle analysis.

C-terminus   End of an amino acid chain; "c" stands for "carboxyl."

Cytoskeleton   A protein-based scaffolding within cells that keeps the cell's shape and enables it to move.

Dendrimers   Macromolecules with a repetitively branched, well-defined structure. Biologically active nanodevices are constructed from dendrimers and used in drug delivery and targeting. Dendrimers are designed and used in the treatment of cancer and other diseases because of their ability to recognize diseased cells, diagnose disease states, deliver drugs, and report locations and outcomes of therapy.

Dimer   An oligomer composed of two monomers.

DNA amplification   The process of quickly reproducing minute quantities of DNA samples for use in genomic analysis.

Dodecamer   An oligomer composed of twelve monomers.

dsDNA/RNA   Double-stranded DNA/RNA.

Dynein   A family of proteins that convert the chemical energy in ATP into motor energy.

Electrophoresis   A method for separating and analyzing macromolecules, such as DNA or RNA.

Electrostatic interaction The force of attraction or repulsion between neighboring particles.

Encapsidation Process by which a virus's nucleic acid is enclosed in a capsid.

Entropy Measures the amount of energy that cannot be used to produce work and how evenly energy is distributed in a system.

Enzyme Macromolecule composed of protein subunits that catalyzes or accelerates a biochemical reaction.

Eukaryote Organism with complex cells or a single cell with a complex structure. They include animals, plants, algae, and fungi. Eukaryotic cells possess many different internal membranes and structures and a cytoskeleton.

F-ATPase The primary enzyme for ATP synthesis.

Folding (chemistry) The process by which a molecule assumes its shape or conformation. In the case of proteins, it refers to the process by which a protein chain assumes its native three-dimensional structure, held together by hydrogen bonds.

Förster or fluorescence resonance energy transfer (FRET) Physical phenomenon of energy transfer between from a donor fluorophore to a nearby acceptor chromophore (or between chromophores). Combined with optical microscopy, it allows one to determine the approach between two molecules to within several nanometers. FRET is also a useful tool for measuring protein–protein interactions, protein–DNA interactions, protein conformational changes, and observing biomotor activity.

FtsK A family of proteins located in eukaryotic cells that is responsible for coordinating the accurate copying of the entire complement of genetic material and the late stages of chromosome segregation with cell division.

Genome An organism's complete set of chromosomes or genetic material, containing all of its hereditary information. Also refers to the complete set of genes in a cell or virus.

Helical Referring to a helix.

Helicase Enzyme whose function is to unpackage an organism's genes. It has an essential role during DNA replication because it separates the double strands into single strands, allowing them to be copied. They are also referred to as motors powered by energy from NTP hydrolysis to play a role in translocation and the unwinding of double-stranded nucleic acid polymers. Helicases are also involved in a number of different cellular and biomolecular processes.

Hexamer An oligomer composed of six monomers.

Hill coefficient Provides a quantitative measure of the cooperativity of ligand binding.

Holliday junction A cross-shaped nucleic acid structure containing four double-stranded arms joined together that forms during genetic recombination.

Holliday junction resolution The repair of breaks in DNA, linking homologous chromosomes.

Homologous (chemistry) A group of compounds with similar chemical makeup and structural features.

Homologous recombination The exchange of genetic material between two similar or identical strands of DNA, which can cause repair in dsDNA breaks and produce new combinations of DNA sequences during meiosis.

Homo-trimer (or homotrimer) A protein composed of three identical polypeptide units.

Hydrolysis The breakdown of chemical bonds by the addition of water.

Icosahedral Consisting of a shape with 20 equilateral triangular faces, similar to a sphere.

Kilodalton (kDa) One thousand daltons (Da), where a dalton is the weight of one hydrogen atom.

Kinesin A protein found in eukaryotic cells that supports several functions including meiosis, mitosis, and intracellular transport of cellular cargo from the center of the cell outward along microtubules.

KOPS (stands for FtsK Orienting Polar Sequences) sequence FtsK-orienting polar sequences that interact with FtsK.

Ligand A biochemical substance that bonds to a biomolecule to form a complex.

Liposome Artificially-formed vesicle that can be used as a vehicle to deliver nutrients and drugs. A number of drugs have liposomal delivery systems.

Lysine A basic amino acid that is a component of most proteins.

Macromolecule *See Polymer*

Magnetic tweezers Instrument used for manipulating and studying the mechanical properties of single molecules by introducing magnetic gradients. One end of the molecule is tethered to a surface; the other is attached to a magnetic bead, which is manipulated via external magnets.

Microtubule Hollow tube that helps maintain a cell's shape and motility, chromosome movement in cell division, and organelle movement.

Minus strand Complement to a plus strand in DNA or RNA that serves as the source for protein code.

Monomer Single atom or molecule that joins with other monomers to form polymers (*mono-* means "one" and *-meros* means "a part") in a process called polymerization. Can be referred to as a "building block."

Motif A short cluster of amino acids or nucleotides that share structural and usually functional similarity. It is linked with a particular function and has a three-dimensional structure.

Myosin A family of proteins that, in combination with actin, is involved in muscle contraction and other kinds of cell movement. There are many different kinds of myosins that are involved in a variety of motor functions.

Nanometer (nm) One billionth of a meter. Expresses measurements on an atomic (related to atoms) scale.

Nanopore Very tiny hole; DNA can pass through it via translocation.

Nucleocytoplasmic large DNA viruses (NCLDV) superfamily A group of viruses with large DNA genomes that replicate in the cytoplasm and partly in the nucleus of eukaryotic cells.

***n*-Fold symmetry** A characteristic that indicates that an object will look the same even after *n* rotations (or folds) or that it has *n* equivalent parts.

Nucleoside triphosphate (NTP) A nucleotide that contains a nitrogenous base, sugar, and chain of three phosphate groups bound to ribose. ATP is a type of NTP.

Nucleoside triphosphate hydrolase (NTPase) An enzyme that catalyzes the hydrolysis of NTP in the same way that ATPase does for ATP.

Nucleic acid Large, complex organic molecule that is a carrier for hereditary information. DNA and RNA are types of nucleic acids.

Nucleoid An area of a prokaryotic cell similar to a nucleus of a eukaryotic cell but is lacking a nuclear membrane.

Nucleotide Chemical compound including a base, a sugar, and one or more phosphates that is the building block of DNA and RNA.

Oligomer A molecular complex consisting of only a few monomers, in contrast to polymers where the number of monomers is unlimited.

Optical tweezers An instrument used for manipulating microscopic objects as small as a single atom by affixing it to a micron-sized glass or polystyrene bead using a highly focused laser beam, creating an optical trap that is able to trap individual particles. Used for studying biological motors and the physical properties of DNA.

Orthologue One or more gene sequences with a similar structure found in different species.

Packaging (in the context of DNA) The process of tightly coiling a DNA molecule to allow it to fit into a cell or virus particle.

Pentamer an oligomer composed of five monomers.

Phi29 (or Φ29) DNA polymerase Comes from the bacteriophage phi29 (which is among the smallest known dsDNA phages); has a number of features that make it suited for use in molecular biology for multiple displacement DNA amplification.

Phosphate A chemical compound composed of phosphorus and oxygen that is a component of nucleic acids.

Photobleaching Alteration of a fluorophore molecule so that it is no longer able to fluoresce.

pico-Newton (pN) A measurement of force equivalent to 10^{-12} Newtons (or one-trillionth of a Newton). A Newton is defined as the force needed to make 1 kilogram of mass accelerate by 1 meter per second.

Phosphate binding loop (P-loop) An ATP binding site present in ATP-driven motors. Also referred to as a protein fold. P-loop NTPases have Walker A (the p-loop proper) and Walker B motifs.

Plus strand Template strand of DNA or RNA containing the instructions for protein building that is paired with a complementary minus strand.

Proton motive force (*pmf*) A process that occurs when the cell membrane becomes energized, promoting the movement of protons across it.

Polymer A chain-like molecular complex consisting of an unspecified, large number of monomers. Sometimes referred to as "macromolecule." Monomers have two carbon atoms joined together to form a carbon double bond, allowing it to make a large chain of polymers.

Polymerase An enzyme that is used to assemble DNA or RNA molecules by synthesizing long chains of DNA or RNA by linking smaller molecular units.

Polymerization The process by which monomers join together to form a polymer through the use of polymerase.

Packaging RNA (pRNA) Hexagonal ring of RNA molecules that the phi29 bacterial virus uses to translocate DNA into the viral procapsid.

Procapsid A capsid containing no viral nucleic acid.

Prohead An immature virus capsid formed in the early stages of self-assembly in some bacteriophages.

Prokaryote An organism that lacks a nuclear membrane or nucleus, mitochondria, and organelles. Instead, they have a single chromosome inside a nucleoid.

Promoter Sequence of DNA where transcription of a gene begins; RNA polymerase binds to it and begins transcription. Also determines the direction of transcription and which strand will be transcribed.

Replication (chemistry) The process in which DNA or RNA is copied to produce two identical molecules.

Rho factor A hexameric helicase that signals to RNA polymerase when to stop transcription.

Ribosome A macromolecular particle containing RNA and other proteins that can be found floating in the cytoplasm. It is responsible for protein formation and synthesis. Protein synthesis takes place within ribosomes. There are typically thousands of ribosomes within a cell.

RNA polymerase An enzyme that synthesizes RNA by following a strand of DNA. It copies a sequence of DNA into an RNA sequence during the transcription process, which it also controls.

Single-molecule sensing A means of measuring, detecting, and monitoring individual molecules and quantification of target species.

SpoIIIE DNA translocase that plays a role during sporulation; also involved in DNA conjugation, segregation, and translocation as well as protein transport.

Secreted phosphoprotein 1 (SPP1) A gene found in bones.

ssDNA/RNA Single-stranded DNA/RNA.

Stoichiometry A branch of science that measures the amount of matter involved in chemical reactions. Used to measure the amount of a substance that will be needed to produce a reaction. Also refers to the number of drug molecules bound to each virion or target related to viral replication.

Substrate The surface on which an organism lives.

Synthase An enzyme that catalyzes the linking of two molecules without an energy source such as ATP.

Terminase Protein involved in DNA packaging. Also referred to as the ATP-powered motor that packs the genetic material inside the bacteriophage capsid. It contains two subunits, the small and large terminase. The small terminase initiates packaging of the viral genome; the large terminase is responsible for translocating DNA.

Tetramer An oligomer composed of four monomers.

Toroidal Ring or donut-shaped.

Transcription The process of forming DNA from RNA during which the information stored in a molecule of DNA is copied into a new RNA molecule, controlled by RNA polymerase.

Translocase A protein or enzyme that facilitates translocation.

Translocation The moving or threading of a molecule across a membrane or through a nanopore.

Trimer An oligomer composed of three monomers.

Vesicle Small, membrane-enclosed sac within a cell that stores or transports substances intracellularly.

Virion A complete virus particle with RNA or DNA at its core and a protein coat. Its function is to deliver the DNA or RNA into the host cell. Serve to protect and deliver the genome and to interact with its host.

Walker A motif (also known as P-loop) A motif in proteins associated with phosphate binding that is known to have a highly conserved three-dimensional structure.

Walker B motif A motif in proteins with a highly-conserved three-dimensional structure that plays a key role in ATPase activity.

Western blot A method used to detect amino acids in a tissue sample

Wild-type The opposite of mutant; refers to substances, organisms, or characteristics that occur naturally or in natural conditions.

Index

Printed and bound by CPI Group (UK) Ltd, Croydon, CR0 4YY

01/11/2024

01782617-0019